WITHDRAWN

*A Cell Biologist's
Guide to Modeling
and Bioinformatics*

THE WILEY BICENTENNIAL–KNOWLEDGE FOR GENERATIONS

*E*ach generation has its unique needs and aspirations. When Charles Wiley first opened his small printing shop in lower Manhattan in 1807, it was a generation of boundless potential searching for an identity. And we were there, helping to define a new American literary tradition. Over half a century later, in the midst of the Second Industrial Revolution, it was a generation focused on building the future. Once again, we were there, supplying the critical scientific, technical, and engineering knowledge that helped frame the world. Throughout the 20th Century, and into the new millennium, nations began to reach out beyond their own borders and a new international community was born. Wiley was there, expanding its operations around the world to enable a global exchange of ideas, opinions, and know-how.

For 200 years, Wiley has been an integral part of each generation's journey, enabling the flow of information and understanding necessary to meet their needs and fulfill their aspirations. Today, bold new technologies are changing the way we live and learn. Wiley will be there, providing you the must-have knowledge you need to imagine new worlds, new possibilities, and new opportunities.

Generations come and go, but you can always count on Wiley to provide you the knowledge you need, when and where you need it!

WILLIAM J. PESCE
PRESIDENT AND CHIEF EXECUTIVE OFFICER

PETER BOOTH WILEY
CHAIRMAN OF THE BOARD

A Cell Biologist's Guide to Modeling and Bioinformatics

Raquell M. Holmes

Center for Computational Science
Boston University
Boston, MA

WILEY-INTERSCIENCE
A JOHN WILEY & SONS, INC., PUBLICATION

Published by John Wiley & Sons, Inc., Hoboken, New Jersey
Published simultaneously in Canada

For general information on our other products and services or for technical support, please
contact our Customer Care Department within the United States at (800) 762-2974,
outside the United States at (317) 572-3993 or fax (317) 572-4002.

Wiley also publishes its books in a variety of electronic formats. Some content that appears in print
may not be available in electronic formats. For more information about Wiley products, visit
our web site at www.wiley.com.

Library of Congress Cataloging-in-Publication Data

Holmes, Raquell M.
   A cell biologist's guide to modeling and bioinformatics/Raquell M.
Holmes.
      p. ; cm.
   Includes index.
   ISBN-13: 978-0-471-16420-3 (cloth)
   ISBN-10: 0-471-16420-8 (cloth)
   1. Computational biology. 2. Bioinformatics. I. Title
   [DNLM: 1. Computational Biology—methods. 2. Cell Cycle—physiology.
   3. Models, Biological. QU 26.5 H767c 2006]
      QH324.2H65 2006
      570'285–dc22

                                                        2006036932

Printed in the United States of America

10 9 8 7 6 5 4 3 2 1

*This book is dedicated to my family, friends, and colleagues,
current and past, who continue to travel with me in the creation
of new meaning and understanding in science and life …*

Contents

Book Overview

WHO THIS BOOK IS FOR

This book is written with the practicing cell biologist and innovative biology educator in mind. This book highlights, through discussion and example, computational and informatics tools currently available for analyzing biological data and modeling cellular processes. It is meant to serve as an introductory text for developing a better understanding of how to use publicly available databases and create computational models. Although the primary intended audience of this book is the researcher in cell biology, the text has been written to be accessible to undergraduate students who have taken or are taking cell biology courses. Undergraduate students were involved throughout the creation of the book as research assistants. They came from multiple disciplines: biology, psychology, physical anthropology, math, chemistry, and biomedical engineering. Their common characteristic was their desire to have a research experience and learn something about computational biology. These students, sophomores to seniors and an occasional graduate student, created models, reviewed text, and refined concepts that are discussed in this text. They have been collaborators in the creation of what I hope will be a useful resource to the cell biology community.

USEFUL SKILLS AND TRAINING

As is the case in most research areas, the more training and exposure to the theory and practice of a methodology, the better prepared one is to use that method appropriately. However, biologists do not need to become computational biologists in order to make use of computational tools. An analogous situation is determining whether to become a molecular biologist in order to use molecular approaches in an area of research. Many cell biologists take advantage of molecular tools (i.e., gene constructs, anti-sense RNA, molecular probes) while investigating aspects of gene transcription or regulation in the context of a larger cellular behavior. A cell biologist does not become a molecular biologist solely by using molecular tools. The cell biologist and the molecular biologist each have a different focus and area of in-depth expertise. It is not required for cell biologists to become computational scientists in order to make use of computational tools. However, it is important to become familiar with the strengths and weaknesses of various computational approaches and to become familiar with the language of modelers and computationalists. By becoming familiar with the language and methods, we are in a better position to develop collaborations and take advantage of the skills of full-time modelers or computational scientists.

We assume that the reader has a working understanding of algebra at the level commonly obtained with completion of high school and early years of undergraduate education. The modeling chapters make the process of modeling obvious and assume that the reader knows little to nothing about the mathematical modeling. The chapters covering sequence alignment and family-domain databases lightly discuss the statistical measures of significance that rely on concepts of frequency and distributions. References are given for primary research and for reviews that provide more in-depth treatments of the statistical underpinnings.

WHAT IS COVERED

This book brings together two topics that are not frequently introduced in the same text: biological database searching (sequence similarity, protein families, and domains) and dynamic models of cellular processes (cell cycle, calcium waves, glycolysis).

The layout of the chapters is designed to progress from information on nucleic acid sequences to proteins and protein functional motifs to cellular behaviors of metabolic pathways, cell division, and calcium dynamics. This progression is from static data and information to simulated behaviors.

Molecular Sequence Database Chapters

The initial chapters are dedicated to the discussion of sequence alignments and the searching of biological databases. These are primarily Web-based, publicly available resources that are popularly used in background research or characterization of novel gene products. The goal of these chapters is to provide the reader with an overview of the essential features (queries, alignment methods, and statistical significance measures) needed to understand and improve searches for sequence similarity, protein families, and putative functional domains.

Modeling Chapters

Each chapter focuses on a basic research topic within cell biology (cell cycle, calcium dynamics, and glycolysis). The goals for each chapter are to take you, the reader, through the creation of a computational model of a presumably familiar biological topic. Each chapter introduces a new simulation tool so that one can experience different simulator interface designs. The tools used in the chapters are by far not the only tools that can be used.

The first half of each chapter discusses the biology and develops a noncomputational model (biological concept map, system statements, and mathematical descriptions) that is independent of the software tools. Thus, one can use the model description in other software applications or in discussions with collaborators.

Each chapter is based on the research publications of experimentalists and modelers. They are designed such that the papers can be read alongside of the chapter. These are additional resources for in-depth investigation of the modeled biological process. The chapters contain detailed introductions to creating the mathematical model based on the knowledge of the biological system. The chapters are written to make evident the process and skills involved in traversing biological and computational models.

TO GET THE MOST OUT OF THIS BOOK

The biological databases are great resources for biological data. They are easily accessed via the Web and contain a wealth of biological data. Overall, the tools are easy to use, requesting as little as a single piece of information to multiple pieces in order to perform a database search. The chapters provide the background information for interpreting search results. To best understand the nature of the data and information that can be retrieved at any of the resources requires going to the sites and running a few searches.

The process of modeling is conceptually easy to understand. However, what appears simplistic is quite challenging and significantly instructive. Modeling is learned by creating or re-creating models, getting stuck, and seeing whether you can work your way out or convince yourself that the model is at a minimum accurately implementing the kinetic or mechanistic relationships between components. To get the most out of the modeling chapters, re-create the models while reading the chapter. The process of transforming conceptual models to formalized relationships with consistent notations and subsequent mathematical representations is an intellectual exercise that advances one's understanding of the biological system even before getting to a computer.

Acknowledgments

Many people and communities of practice have contributed to the writing of "A Cell Biologist's Guide to Modeling and Bioinformatics". The book benefits from the knowledge and expertise of scientists in computational cell biology, educators in high performance computing and biology, as well as, performatory and project-based educational approaches developed by practitioners in human development and biology education reform. This diverse grouping includes the members of the Center for Computational Science; partners of the Education Outreach and Training Partnership for Computational Science (EOT-PACI, 1997–2005); students of Boston University; researchers within the growing Computational Cell Biology community; and colleagues and trainers from East Side Institute for Group and Short Term Psychotherapy and the BioQUEST Curriculum Consortium.

There are particular teachers within these communities that I wish to acknowledge. Roscoe C. Giles at Boston University for supporting me as scientist, educator and colleague; Tom Bartol and Joel Stiles, developers of MCell, who share tremendous knowledge in concise, egalitarian conversations; Herbert Sauro who excitedly shares new resources and engages learners; and Ann Cowan, Les Loew, Jim Schaff, and Ion Moraru, consistent organizers of new directions in quantitative cell biology. The more than 32 partners in EOT-PACI are too many to list. My thanks to all those who engaged in practical, detailed conversations of kinetics, regression analysis, stochastic and deterministic methods. Special thanks go to the members of the BioQUEST Curriculum Consortium who were the first group in which I found the shared spirit of community organizers, biologists, educators, and technology enthusiasts.

This book was made possible by the funding support of the NSF through EOT-PACI and Engaging People In Cyber infrastructure (EPIC) and the human support of students, predominately undergraduates, who took on learning something they had not known existed—computational biology—in order to advance their own academic development. The students have been my lab collaborators. They include Andrea Allen, Mona Aoude, Alisa Chalmers, Walton Chiu, James Diamond, Amy Hyme, Justin Lui, Susan Ma, Andrea Matthews, Deidre Morrison, Andrew Nocon, Le Nguyen, Maryam Qureshi, Nathalie Redziniak, Katrina Schrode, Dameon Shaw, Aya Takakayu, and Geetu Tuteja.

Chapter *1*

Introduction

1.1 A NEW TALE OR SAME STORY, DIFFERENT DAY?

In this day and age of informatic and 'omic efforts within biology, experimentalists are challenged to exploit information systems and computational models within their research. Computational tools that are already commonly used include image analysis software, structural modeling programs, and sequence alignment tools. Mellman and Misteli (2003) suggest that it is time for computation to become recognized as a tool on a par with molecular tools in cell biology research. At the molecular level, one rarely questions if Web-based sequence alignment and database tools should be used but rather asks if they are being used efficiently and accurately to produce trustworthy results. At the cellular level, informatics tools are helpful in identifying putative molecular components and functions. However, the dynamic behaviors of cellular systems require the development of computational models (mathematical models). Mathematical models are becoming increasingly visible in the cell biology literature, and yet the methods of creating such models are less obvious to many of us trained as experimental cell biologists.

A number of opinion pieces have been published in journals commonly read by cell biologists—*Nature*, *Science*, *Journal of Cell Biology*, and *Cell*—that engage the question of what cell biologists are to do with the now-existent parts lists generated from the Human Genome Project and ongoing genomic and proteomic work (Bray, 1997; Hartwell *et al.*, 1999). Does having such an extensive catalogue of molecular data change our understanding of the nature of biology? Hartwell argued that biology may be better understood as modular. Others have described biology as an information science, relating to the implementation of instructions encoded in DNA and RNA and operationalized by protein, carbohydrate, and lipid machinery. These perspectives draw on lessons and advantages gained from engineering principles and information sciences. Whether these perspectives are helpful frameworks for further understanding biology has yet to be determined. A common theme within these perspectives is an increased focus on the complex and formalized relationships between biological factors. These include evolutionary relationships that provide evidence for gene function and protein and network interactions that describe molecular circuitry underlying cellular processes and formalized kinetic relationships in the form of rate equations used to simulate dynamic behaviors.

Despite the recent fervor and attention to the benefits of computing in biology, the use of computational approaches in biology has existed for years although not always in association with computers. What we now commonly refer to as computational models appear historically as mathematical and theoretical models. As such, computational models in biology can be found as early as 1952 when Turing hypothesized short-range action, long-range inhibitor reaction diffusion as an explanation of pattern formation. This theory was later applied to shells, cheetahs, and drosophila (Nagorcka and Mooney, 1992). Membrane physiologists have used mathematical models to characterize membrane proteins such as pumps and channels; and the use of mathematical models and computation has led to better understanding of actin polymer dynamics, muscle contraction, and drosophila development (Julian, 1969; Pollard, 1986; Wachsstock and Pollard, 1994). Although computations have been part of our work, these aspects of our research have not been explicitly discussed as often by us as experimentalists. We are most familiar with mathematics in terms of probability associated with graphs and tables to demonstrate that our experimental results are not likely due to chance.

Recently, standard journals read by cell biologists—*Science, Nature, Journal of Cell Biology*—are publishing more papers that include computational analysis. Reviews discuss the importance of modeling and simulation in understanding the dynamics of cellular systems and how we approach biological research (Hartwell *et al.*, 1999). Research papers describe modeling results in an effort to better understand specific biological systems. Vesicular transport, membrane ruffling, and cell adhesion are just a few examples (Hirschberg *et al.*, 1998; Waterman-Storer and Danuser, 2002; Lee *et al.*, 2003). Within these papers, there is either an implicit or explicit reference to the importance of modeling.

1.2   COMPUTATIONAL BIOLOGY

Computational biology is a broad discipline, as broad as the numerous fields of biology and methods of computation. In its simplest description, computational biology is the use of computers and mathematics to solve problems within biology. Computation involves applying known and hypothesized relationships in mathematical form to the description of phenomena. The use of computational methods in biological research has been referred to in many ways, typically dependent on the biological focus and computational method of the speaker. Overall, the terms refer to the development and use of mathematical descriptions of a working hypothesis.

Bioinformatics   Algorithms and database designs focused on molecular data or information management, sometimes including protein folding efforts.

Quantitative Biology   Used in reference to various biological scales (e.g., molecular, cellular, tissue, etc.) and refers primarily to measuring quantities of biological factors. The quantitative data is then subjected to informatics efforts for cataloguing and is available for numerical modeling.

Systems Biology   Complex systems exist within and across multiple biological scales. Systems biology has been defined as the determination of the components of these biological systems as well as the simulation of system behaviors with kinetic models in a given scale (Henry, 2003).

Computational Cell Biology   Mathematical models of cellular systems including molecular motors, vesicle transport, cell signaling, and actin dynamics.

Bioinformatics is an aspect of computational biology that uses statistical approaches to model biological relationships (e.g., evolutionary traits, structural and gene regulatory networks). Broadly, bioinformaticians work with and develop computational methods and information management systems to discover biological principles. In practice, bioinformatics has developed the tools by which researchers can aggregate background information on already characterized genes and proteins as well as tools to predict genetic and biochemical networks across species.

A key characteristic of the 'omic efforts is the need for high-throughput methods of generating data. The development of bioinformatics from genomics was largely due to the ability to mass-produce nucleic acid data for DNA mapping, gene identification, sequencing, and expression mapping. The great success of the genomics effort has led to the search for methods of generating large reproducible, reliable, and biologically relevant data sets for proteomics, metabolomics, cellomics, and physiomics. Researchers in these areas focus on engineering high-throughput data methods and computational tools to mine and analyze the data. 'Omic research enables discovery science, where statistical models are used to identify patterns of biological significance from the data.

Molecular sequence data and computational tools are used to develop arguments of molecular identity, homology, and function that are based on evolutionary relationships and the presence of domain motifs/signatures, and 3D structures. DNA sequences of known genomes are used to search for and predict homologous or orthologous biological functions in species where they have yet to be identified through experimentation. Appropriate annotations are sought for genes in the databases, and researchers attempt to infer function based on sequence and expression data. This includes determining participation in genetic or metabolic networks. Computational tools are employed to (1) draw comparisons between DNA sequences, gene structures, and determine possible evolutionary relationships, (2) predict biochemical properties of proteins (i.e., protein folds, protein binding sites, and posttranslational modifications), and (3) predict the functional role of genes in cellular or physiological processes.

Molecular sequence databases are useful for determining what is known experimentally or predicted computationally about the molecular components of biological systems and their biological function. Sequence similarity searches are used to infer identity and biochemical properties of a novel cDNA clone based on already characterized genes and proteins. Protein family and domain databases use sequence and structure data to construct family assignments or characteristic patterns for functional domains. Databases of protein-protein interactions provide evidence for *in vivo* interactions and participation in biochemical pathways. Protein interaction data, microarrays, and protein profiles are subjected to cluster analyses to infer coregulated genes. Genomic comparisons are used to identify conserved pathways between species.

The challenge that we face as experimentalists is how to isolate relevant information from the large pool of data.

Which data resource has the type of data one is looking for?

Is the data within the database trustworthy?

How does one know if one's search results are statistically and, importantly, biologically significant?

In the following chapters on sequence alignment and protein family and domain databases, we provide initial insights to answer these questions. More in-depth answers and direction can be found in books and resources dedicated to informatics discussion and training. The treatments in this book provide background information and directions that can be useful to researchers as they engage with more commonly used Web-based molecular research tools.

1.3 MODELING AND SIMULATION

As the use of information systems increases in biological research, allowing us both to catalogue and search large amounts of data, biologists are faced with a new opportunity to study complex relationships that were previously not feasible. We are challenged to create models as a means of (1) making explicit the understood functional relationship between biological entities (i.e., protein, nucleic acid, or organelle), (2) finding fault with a hypothesis, and (3) providing colleagues an additional means of testing our results.

Experimental biologists typically use conceptual models to illustrate current hypotheses or understandings of cellular systems. These explanations are usually provided in the form of illustrations, information flow charts, and diagrams. These diagrams replace the need for developing paragraphs of text to define the relationships among biological factors within any system. However, it is impossible to use these diagrams to test the hypothesis. In a computational model of dynamic systems, the relationships among factors in a diagram are made explicit by defining the relationships in terms of rates, quantities, or state changes. This transforms conceptual models into *working hypotheses.*

The phrase *mathematical description* can invoke images of a series of equations consisting of unfamiliar symbols or incomprehensible numbers. It is helpful to state here that generating a mathematical description is not necessarily about numbers or exact quantities but rather about formalizing the relationships between biological objects such that the relationship itself and the product of the relationship can be tested. Developing models therefore is the mathematical assertion of a hypothesis about our experimental system in a testable form. Transforming our hypothesis into explicit relationships and assumptions is a rigorous reflective process that helps to expose missing components and inconsistencies.

The topics discussed in this book might be considered the basis for computational cell or systems biology. Molecular databases provide information on cellular components, and mathematical models provide a means of studying dynamic properties of biochemical reactions. The computational methods differ for informatics and dynamics modeling. For instance, sequence alignment and search algorithms use statistical models to determine the significance of an alignment. This is distinct from the hidden Markov models used to calculate protein domain motifs, graphical models for network inference, differential equations used to simulate dynamical systems, or Monte Carlo simulations of stochastic processes.

1.3.1   From Databases to Dynamics: A View of Network and Pathways

The terms *cell signaling*, *pathways*, and *networks* are used to describe a series of protein interactions that are the basis of dynamic, observable cellular behaviors. Intricate sets of interactions have been created from the examination of interactions between subsets of proteins. For example, we investigate growth factor stimulation of a signaling pathway by examining changes in the state of one or more downstream molecules (e.g., cell adhesion activation of PI3 kinase; Epidermal Growth Factor (EGF)-triggered phosphorylation of Mitogen Activated Protein Kinase (MAPK)). Changes in the phosphorylation state or localization of a protein can serve as the biomarker for the activation of the pathway. This biomarker is then used to identify additional players. The reduction and subsequent reconstruction of the complex system of interactions is ideal for experimentation. However, the properties of the system are not reflected in the discussion of the details or parts of the system but rather only in the discussion of the system as a whole (*Nature* 403: 345–346). The network is a view of the global aspects of the system. A distinction between pathways and networks then is the level of abstraction.

To infer of gene and biochemical networks, researchers use information science, statistics, and graph theory to integrate data and elucidate complex biological relationships. They take advantage of data from genomic and gene and protein sequence databases to map newly identified sequences onto preexisting, already established networks or to predict new interaction networks. Network and pathway databases like molecular sequence databases catalogue information. They differ from the sequence database by focusing on the interactions between genes and proteins. Kyoto Encyclopedia of Genes and Genomes (KEGG) maps known metabolic pathways in yeast and computes similar pathways in other species (Table 1.1; Goto *et al.*, 1997; Ogata *et al.*, 1999; Kanehisa *et al.*, 2006). EcoCyc contains *Escherichia coli* metabolic and some signaling pathways and is the metabolism template for predicting metabolic pathways in other species (Karp *et al.*, 1999). Biomolecular Interactions Network Database (BIND), now encompassed by BondPlus and IntAct developed by the European Bioinformatics Institute, catalogues protein-protein interactions (Hermjakob *et al.*, 2004; Unleashed Informatics, Ltd., 2006). Collectively, these databases gather and organize data on networks and pathways, as well as provide computational tools that primarily use binary relationships to predict metabolic pathways in species where they have yet to be confirmed experimentally.

When examining networks and pathways, there are a series of questions that can be posed with which computational and bioinformatics tools can help. These include: What genes or proteins are involved? What are their functions? What other pathways or networks are they involved in? Is the associated cellular behavior dependent on the concentration or location of the factor?

TABLE 1.1   Gene and Protein Interaction—Network Databases

Database	Type of Data	Primary Species	Data Sources
BIND	Protein interactions	Yeast	
KEGG	Gene and metabolic networks	Yeast	WIT, LIGAND, Enzyme handbook, Japanese catalogue
EcoCyc	Metabolic pathways	*E. coli*, coding DNA only	Genbank, Enzyme, primary literature
CSNDB	Cell signaling	Human	Transfac, journal literature

TABLE 1.2 Computational Resources for Addressing Biological Questions

Question	Computational Resources
What are the functions of the proteins?	Molecular sequence, family, and domain databases
What proteins are involved in a given pathway?	Genetic and metabolic pathway database
What is the dynamic behavior of protein interactions?	Computational simulation
How are dynamics of cellular behaviors affected by changes in molecular concentrations and kinetics?	Computational simulation

These questions can be addressed by different bioinformatics and computational resources (Table 1.2). The database tools and graphical maps are helpful toward understanding the protein components and interaction flows in a pathway, and mathematical modeling tools enable us to examine and better understand the behavior of the pathways and networks.

Mathematical models take many forms. The models discussed in this book are systems of differential equations also known as continuous or population models. These equations are solved numerically by providing numerical values for concentrations, rates of reactions, diffusion rates, and binding constants. Quantitative data for some of these values exists in literature and databases, however many are missing for a large fraction of known proteins and enzymes. National Institutes of Health—funded efforts of the Alliance for Cellular Signaling (AFCS) and the National Technology Centers for Networks and Pathways have focused on the development and use of methods to obtain this quantitative data. The premise is that by obtaining quantitative data within biological systems, it will be possible to model the dynamics of cellular networks and pathways and thus predict the behavior of a system.

BIBLIOGRAPHY

Alfarano C, Andrade CE, Anthony K, *et al.* (2005). The Biomolecular Interaction Network Database and related tools 2005 update. *Nucleic Acids Research* 33(Database issue):D418–D424.

Anonymous (2000). Can biological phenomena be understood by humans? *Nature* 403(6768):345.

Baldi P, Brunak S (1998). *Bioinformatics: The Machine Learning Approach.* Cambridge: MIT Press; 351p.

Bray D (1997). Reductionism for biochemists: how to survive the protein jungle. *Trends in Biochemical Sciences* 22(9):325–326.

Goto S, Bono H, Ogata H, *et al.* (1997). Organizing and computing metabolic pathway data in terms of binary relations. *Pacific Symposium on Biocomputing* 2:175–186.

Hartwell LH, Hopfield JJ, Leibler S, *et al.* (1999). From molecular to modular cell biology. *Nature* 402:C47–C52.

Henry CM (2003). Systems biology. *Chemical and Engineering News* 81(20):45–55.

Hermjakob H, Montecchi-Palazzi L, Lewington C, *et al.* (2004). IntAct: an open source molecular interaction database. *Nucleic Acids Research* 32:D452–D455.

Hirschberg K, Miller CM, Ellenberg J, *et al.* (1998). Kinetic analysis of secretory protein traffic and characterization of Golgi to plasma membrane transport intermediates in living cells. *Journal of Cell Biology* 143:1485–1503.

Julian FJ (1969). Activation in a skeletal muscle contraction model with a modification for insect fibrillar muscle. *Biophysical Journal* 9:547–570.

Kanehisa M, Goto S, Hattori M, *et al.* (2006). From genomics to chemical genomics: new developments in KEGG. *Nucleic Acids Research* 34:D354–357.

Karp PD, Riley M, Paley SM, *et al.* (1999). EcoCyc: an encyclopedia of Escherichia coli genes and metabolism. *Nucleic Acids Research* 27:55–58.

Lee K, Dinner AR, Tu C, *et al.* (2003). The immunological synapse balances T cell receptor signaling and degradation. *Science* 302:1218–1222.

Lippincott-Schwartz J, Snapp E, Kenworthy A (2001). Studying protein dynamics in living cells. *Nature Reviews* 2:444–456.

Mellman I, Misteli T (2003). Computational cell biology. *Journal of Cell Biology* 161:463–464.

Nagorcka BN, Mooney JR (1992). From stripes to spots: prepatterns which can be produced in the skin by a reaction-diffusion system. *IMA Journal of Mathematics Applied in Medicine and Biology* 9(4):249–267.

Ogata H, Goto S, Sato K, *et al.* (1999). KEGG: Kyoto Encyclopedia of Genes and Genomes. *Nucleic Acids Research* 27:29–34.

Pollard TD (1986). Mechanism of actin filament self-assembly and regulation of the process by actin-binding proteins. *Biophysiology Journal* 49:149–151.

Regev A, Shapiro E (2002). Cellular abstractions: cells as computation. *Nature* 419:343.

Rives AW, Galitski T (2003). Modular organization of cellular networks. *Proceedings of the National Academy of Sciences USA* 100(3):1128–1133.

Setubal JC (1997). *Introduction to Computational Molecular Biology*. Boston: PWS Publishing; 296p.

Sibley CG (1997). Proteins and DNA in systematic biology. *Trends in Biochemical Sciences* 22(9):364–367.

Takai-Igarashi T, Kaminuma T (1998). A pathway finding system for the cell signaling networks database. *In Silico Biology* 1:129–146.

Takai-Igarashi T, Nadaoka Y, Kaminuma T (1998). A database for cell signaling networks. *Journal of Computational Biology* 5(4):747–754.

Tass PA (1999). *Phase Resetting in Medicine and Biology: Stochastic Modelling and Data Analysis*. New York: Springer-Verlag; 329p.

Turing AM (1952). The chemical basis of morphogenesis. *Philosophical Transactions of the Royal Society (B)* 237:37–72.

Unleashed Informatics Ltd. BOND. Available at http://bond.unleashedinformatics.com/index.jsp?pg=0.

Wachsstock DH, Pollard TD (1994). Transient state kinetics tutorial using the kinetics simulation program, KINSIM. *Biophysical Journal* 67:1260–1273.

Waterman-Storer CM, Danuser G (2002). New directions for fluorescent speckle microscopy. *Current Biology* 12:R633–R640.

Chapter 2

Sequence Alignments and Database Searches

2.1 PURPOSE OF THIS CHAPTER

The goal of this chapter is to provide novice users of publicly available sequence databases with sufficient understanding to perform and interpret sequence similarity searches in biological databases. By performing sequence similarity searches against biological databases, newly identified gene sequences can be used to find putative biological relationships. In order to achieve this goal, we must have an understanding of the types of information contained within sequence databases and their records, how sequence alignments are obtained, and how to interpret the results of such searches.

For the purposes of this chapter, we highlight the molecular data types in biological sequence databases, search alignment methods and parameter settings, and the interpretation of statistical values associated with the results. These factors directly affect the researcher's ability to retrieve and evaluate data on sequence similarities and homologies. The examples in the chapter are based on searches using the databases provided by the National Computational Biology Institute (NCBI). Tutorials with step-by-step instructions on the use of NCBI search tools are available from NCBI.

2.2 BIOLOGICAL INTRODUCTION: INFERRING HOMOLOGY

Our basic understanding of cellular proteins is that structure determines function. The ability of a protein to serve as cofactor, enzyme, catalyst, or storage molecule is a property of its structure. The structure when it comes to proteins is determined by sequence, folding, and posttranslational processing. The protein sequence is determined by genes and genomic sequences. It is commonly understood that sequences and structures that

TABLE 2.1   Common Molecular Questions Addressed at Sequence Databases

Common Questions	Approaches
Is my sequence unique?	Sequence comparisons
What biochemical properties could it have?	Sequence comparisons, predicted biochemical characteristics
What other proteins or genes is mine related to?	Sequence comparisons, predicted motif and domain characteristics

resemble one another serve similar functions. Sequences resemble one another by having similar sequences. Stretches of nucleotides or amino acids have the same or similar residues in the same order. From these similarities or lack thereof, biologists infer biological relationships. In terms of evolution, we categorize genes as homologous, paralogous, or orthologues. In terms of cellular function, we use similarities and common characteristics to infer putative cellular roles and activities.

Detail 2.1

Twenty-five percent identity over a stretch of 100 amino acids can be considered good evidence of common ancestry between two sequences.

Sequence information about genes and proteins is also used to construct molecular tools. We want to know the sequence of a specific protein or gene in order to design primers or probes for experiments such as Reverse Transcription-Polymerase Chain Reaction (RT-PCR), RNA inhibition (RNAi), *in vivo* hybridization, Green Fluorescent Protein (GFP) constructs, and so forth. Regardless of the intention, ideology, or end use of the data, researchers are finding biological sequence data and the ability to search annotated sequence databases useful to address common questions about relationships, properties, and identity (Table 2.1).

Search strategies for identifying related DNA and amino acid sequences are built on evolutionary principles and in the most simplistic description engage in a series of sequence alignments to map degrees of identity between sequences from which one infers the degree of homology.

1. Sequence alignment is primary method of comparing sequences.
2. Molecular evolution looks at the conservation of nucleotides or amino acids within sequence or subregions.
3. Homology is an inference based on degrees of similarity.

2.3   SEARCHING FOR SIMILAR SEQUENCES

The goal in creating a search strategy is to optimize the retrieval of sequences related to the sequence of interest. To create successful, efficient searches, it is helpful for us to understand how search methods retrieve records. Searching biological databases for related sequences requires selecting a query sequence(s), a method for searching, and database(s) to be searched (Fig. 2.1). By varying choices for each component of the search, multiple

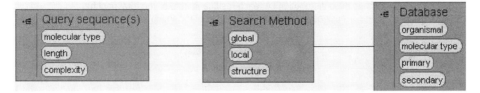

Figure 2.1 *Concept map of key elements used in database search for sequence similarity.*

unique searches can be constructed, and the construction of the search (initial sequence, method, and database) determines the statistical and biological significance of the search results. In the next pages, we discuss characteristics of query sequences, obtaining a sequence from NCBI database records, and use of the sequence in sequence alignment searches against databases.

2.3.1 Query Sequences

2.3.1.1 Types of Query Sequences A query sequence is the nucleotide or amino acid sequence used to search a database and create pairwise sequence alignments. The statistical and biological significance of the sequence alignment and search results is determined by all three factors of the search: query, algorithm, and database. The type of query sequence and purpose of our search direct our subsequent choices of alignment and search algorithm as well as database. The broadest classification of a query sequence is based on its molecular nature as nucleic or amino acid. These are inherently different starting points from a biological and statistical perspective. There are many biological processes between chromosomal DNA to active proteins (e.g., chromatin folding, transcription start frames, exon excision, mRNA translation, etc.). Genomic or even mRNA sequences do not definitively determine the protein sequence of a cellularly active protein. Using protein sequences over nucleic acids when looking for related proteins removes ambiguities that can arise from alternate processing including translational wobble or alternate codon usage.

Looking at DNA and protein from a statistical perspective, we can begin with their molecular alphabets. Nucleic acid sequences are composed of four bases—A, T, G, C—whereas proteins are composed of 20 amino acids. The chance of finding similarities between evolutionarily unrelated sequences is greater in nucleic sequences than in amino acid sequences, just as it would be more likely to identify similarities between random words in a 10-letter alphabet than it is in a 26-letter alphabet (Fig. 2.2). Thus, it is more likely to find stretches of identical sequences within a pairwise DNA-DNA comparison than a protein-protein comparison by chance. The increased likelihood of finding aligned sequences by chance decrease the statistical significance of the alignment. Direct comparison of the ability of DNA versus protein sequence searches to detect sequence similarity have shown that DNA searches return fewer significant matches (Pearson, 1995, 1996, 2000).

Finding quality alignments is highly dependent on the method of scoring residue pairs between two sequences. We know empirically and from evolutionary principles that substitutions occur within sequences that result in related but not identical gene and protein sequences. Bioinformaticians have developed scoring mechanisms—substitution scoring matrices that we discuss in more detail in a later section—that exploit the variety of substitutions possible in nucleic and amino acids. Because of the substitution matrices, more

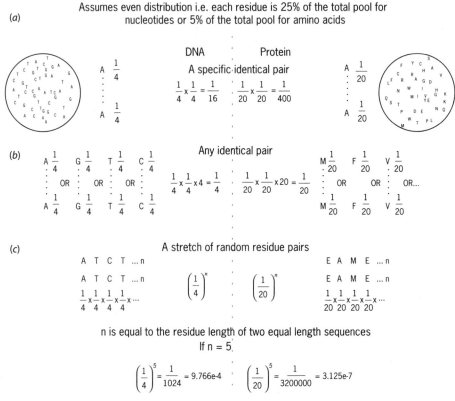

Figure 2.2 *An early model of the probability of alignment of two random sequences. Mirror represen-*
tation of the probabilities for obtaining identical pairs by chance for nucleic acids (left) and amino acids
(right) are shown. (a) The likelihood of selecting a specific identical pair. (b) The probability of having any
matching pair is 1/4 (0.25) for nucleotides and 1/20 (0.05) for amino acids. (c) It follows that the prob-
ability of a random stretch of 5 matching pairs is the probability of any identical pair raised to the power of
the number of residue pairs. In this example, we can see that the likelihood of obtaining matching pairs in
a random nucleotide sequence is three fold greater than in an amino acid sequence of the same length.

sensitive sequence alignments can be made with protein than with nucleic acid queries. If a
researcher only has RNA or DNA sequence data, translational tools can be used to predict
putative protein sequences for genomic, gene, and mRNA sequences (States and Botstein,
1991). This allows one to make use of the substitution scoring schemes while searching for
related sequences.

Highlight 2.1

When possible, use amino acid sequence instead of nucleic acid sequences to search
sequence databases for related sequences.

Other characteristics of the query sequence that are important to take into consideration
include its length and complexity. The length of a query sequence affects the ability of the

search algorithm to identify statistically significant matches within a given database. Short sequences (<20 to 28 nucleic acids; <15 amino acids) are more likely to align by chance within a larger sequence or across many sequences in a database.

Detail 2.2
Fifty percent identity in a 20- to 40-amino-acid region frequently occurs by chance.

Yet, shorter sequences may represent conserved functional motifs such as phosphorylation sites, calcium binding domains, nuclear localization signals, and so forth. To obtain statistically significant matches for these biologically relevant motifs, additional alignment methods and specialized secondary databases have been developed. These include protein family and domain databases, which we discuss in the next chapter.

2.3.1.2 Keyword Searches for Individual Sequence Records
For pairwise alignments, aligning two sequences, we start with a query sequence. If we have recently isolated a sequence, it becomes the material for searching the databases. We may otherwise need to retrieve the molecular sequence of our protein or gene from a database. Ideally, to retrieve a sequence from the public databases, one uses the unique identifier for the sequence. Unique identifiers function similar to Social Security numbers in that they are unique values that correspond with specific individuals. In the case of sequence databases, the unique identifiers are typically strings of letters and numbers, and the individuals are sequences and sequence records. A separate unique identifier is given to a database record and the sequence within the record. This makes it possible to update database records without changing the identifier for the sequence itself. Each record has fields for the record and sequence ID in addition to data fields containing the sequence, publications, sources, and sequence characteristics (coding regions, putative domains). Sequence IDs can be found in articles when submission of the sequence is required for publication in a journal.

In the absence of having the sequence identifier, a common starting point in all Web-based searches is the use of the keyword search. Keyword searches are based on matching query word(s) to one or more words in the data fields of a record. Records contain multiple text fields that may contain the search term. This increases the likelihood of obtaining a search result (match) but also increases the probability of returning records of little or no interest. For example, an unrestricted key word search for "IP3 receptor" will retrieve records with the term(s) in a publication title yet no sequence data for the IP3 receptor.

Keywords are rarely unique to one's protein or gene of interest. The keywords "IP3 receptor" can be parsed by the search engine into IP3 OR receptor, IP3 AND receptor, or IP3 receptor. If the database contains many receptor sequences, but none for the IP3 receptor, the first two sets of search queries could return results, whereas the third would not. As well, some text search algorithms and databases employ synonyms that will find "InsP3" as a match to IP3. Without the synonym matching, the term IP3 would miss records with InsP3.

The number of records retrieved in a keyword search can be quite large. These searches are likely to find unrelated records due to matches in publication fields where the search term is present in the abstract, title, or document text; or matches from the database using a synonym in the search instead of the original term (i.e., "IP3" may be converted to "IP").

Highlight 2.2

As shown in Table 2.2.1, NCBI uses the term *accession number* to refer to a record identifier. In contrast, the term *accession number* is used in Protein Information Resource (PIR) and SwissProt to refer to the sequence identifiers. Regardless whether referring to record or sequence, accession numbers are the most direct means of retrieving information about a particular sequence in a database. Each database has its own naming convention for assigning IDs (Table 2.2.2). Nucleotide accession numbers from the primary databases begin with a single letter followed by five numbers or two letters followed by six numbers. An accession number for a protein sequence translated from the primary nucleotide databases begins with three letters followed by five digits.

HIGHLIGHT TABLE 2.2.1   Frequently Used Names for Unique Identifiers

Molecule: MAP Kinase Kinase2, rat

Database	Nomenclature	Identifier	What's identified
NCBI	Accession number	AAA41620.1	Protein record
NCBI	Gene identifier (gi)	GI:349545	Gene sequence
NCBI	Accession number	L14936	Gene record: mRNA
NCBI	Gene identifier (gi)	gi:349544	Gene sequence
PIR	Entry number	A48081	Protein record: sequence and annotations
PIR	Accession number	A48081	Protein sequence(s)
		S38376	
		S38301	
		S32412	

HIGHLIGHT TABLE 2.2.2   Accession Number Naming Conventions for Records

Accession Number	Record Type (Not Sequence)	Databases
U12345	Nucleotide	GenBank/EMBL/DDBJ
AY11123456	Nucleotide	GenBank/EMBL/DDBJ
NM_123456	Nucleotide: mRNA	RefSeq
NC_123456; NT_123456	Nucleotide: complete genome or chromosome	RefSeq
NG_12345	Genomic region	RefSeq
AAA12345	Translated protein sequence from nucleotide	GenPept
O#####; P#LLL#; Q##L##; P###L#	Protein sequence; 1 [O,P,Q]; 2 [0–9]; 3 [A–Z,0–9]; 4 [A–Z,0–9]; 5 [A–Z,0–9]; 6 [0–9]	SwissProt
123456L; 1234567L	Protein Sequence	PRF
NP_123456	Protein Sequence	RefSeq

Narrowing the search terms and search space to obtain fewer returned records is one means of increasing the likelihood of identifying a record of interest. For example in NCBI, keyword searches can be limited to records for a single species (drosophila, yeast, mouse, arabidopsis), molecular type (cDNA, EST, genomic, protein), or specific record

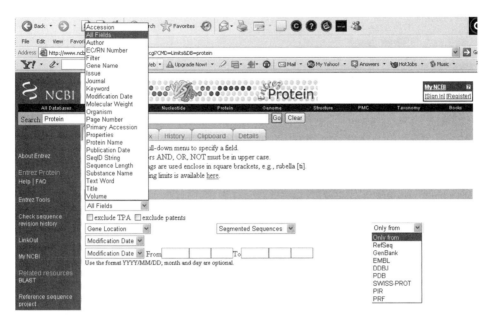

Figure 2.3 Limits can be applied to a NCBI keyword search through drop-down menus, seen on the left and right sides, provided on the "Limits" page.

fields (Fig. 2.3). NCBI currently lets you restrict searches to one of 23 record fields, 3 gene locations (DNA/RNA, chloroplast, or mitochondria), or one of 8 specific databases.

Keyword search results are returned at NCBI as record summaries. This typically includes the *accession number*, a *gene identifier*, and *definition* or *title* of the sequence (Fig. 2.4). The definition line commonly mentions the organism that the sequence is from, the name of the gene product if known, and type of sequence (mRNA, chromosomal). Together the accession number and definition line may provide sufficient information for one to determine if the record contains the desired sequence. However, because annotation errors for putative gene and protein sequence exist, it is important to use more than the title as criteria for success. To determine if the record contains the sequence of interest, use the sequence in a default BLAST (basic local alignment search tool) alignment search. An appropriate sequence will return significant alignments to other related sequences.

2.3.2 Sequence Alignment and Database Searching Methods

In contrast with the keyword search, sequence alignment searches only look in sequence data fields for records of interest. Stated simplistically, search methods engage in a series of sequence alignments to determine degrees of similarity between sequences and then return a list of matched sequences to the user. Multiple methods for sequence alignment have been developed, and more are likely to be created. The key steps to performing automated sequence alignment are align sequences; score the alignment; and, when searching a database, rank and return significant alignment results. The alignment methods that we describe next are based on dynamic programming, and their results are evaluated statistically based on the distribution of related and unrelated (random) sequences. Unless otherwise stated, the principles discussed below apply equally to nucleic and amino acid sequences.

Figure 2.4 An example of a full NCBI record: field names are listed on the left and field contents to the right. NCBI record summaries list a subset of fields, highlighted and boxed in the record example shown. Circled in the record are the organism, gene and sequence type listed in the DEFINITION field.

2.3.2.1 Alignment Algorithms

Determining sequence similarity through sequence alignments is a familiar concept and practice to biologists. Manually, we examine two or more sequences for similar residue patterns, match up identical residues, decide qualitatively whether they are aligned well, and determine statistically how identical or similar the sequences are. The automation of this process requires a computer-based method to line sequences up against one another and a scoring method for evaluating the success of this alignment in terms of similarity or identity. Two methods for accomplishing this automated sequence alignment process are global or local alignment algorithms. Both make use of dynamic programming.

Global Alignments The global alignment method created by Needelman and Wunsch (1970) aligns the entirety of two sequences, comparing the query sequence (A) to the target sequence (B). The goal is to obtain the greatest number of matched residues over the full length of the sequence including the sequence ends. Dynamic programming involves creating a matrix with a sequence on each axis. Every possible alignment between the sequences is explored by calculating an alignment score at each vertices in the matrix.

Assigning values to residue pairs to generate an alignment score is fundamental to sequence alignment and sequence search algorithms. To determine the best alignment, the computer needs a method for scoring alignment. Each nucleotide location on the matrix is examined for its identical match (A-A versus A-T; L-L versus L-G). Each match or mismatch is given a score. A linear path through the matrix that gives the greatest number of matches and subsequently the highest score is the optimal alignment (Fig. 2.5).

(*a*) Scoring scheme: gaps (G), –7; match, 5; mismatch, –4

$$(b) \quad {}^{aS}_{TT} = \max \begin{cases} {}^{aS}_{UL} + {}^{pS}_{TT} & = 5 + 5 = 10 \\ {}^{aS}_{L} + G & = -2 + -7 = -9 \\ {}^{aS}_{A} + G & = -2 + -7 = -9 \end{cases}$$

(*c*)

	A pS aS	T pS aS	C pS aS	G pS aS
A	AA = 5 5	AT = –4 –2	AC = –4 –4	AG = –4 –4
T	TA = –4 –2	TT = 5 10	TC = –4 3	TG = –4 –4
C	CA = –4 –4	CT = –4 3	CC = 5 15	CG = –4 8
G	GA = –4 –4	GT = –4 –4	GC = –4 8	GG = 5 20

Figure 2.5 *(a) The scoring scheme for comparing two sequences specifies values for gaps, matches, and mismatches between residues. (b) The equation for determining the alignment score (aS) is the maximum score of three relationships: the alignment score of the upper left cell on the diagonal plus the position score, the alignment score of the left cell plus the gap score, or the alignment score from the above cell plus the gap score. (c) To illustrate how the scoring matrix is populated, a simple matrix example is provided with nucleotide residues A, T, C, and G. Each matrix cell is shown with the paired residues (e.g. AA), the position score for the pair (=5) and the alignment score (bottom left). Matrix origin is top left. The position score (pS) is the score for the residue pair (e.g. AA, AT) at any given matrix position. The alignment score of the initial pair participates in the determination of alignment scores for neighboring cells as indicated by the arrows. The alignment score for the residue pair TT (shaded box) is 10 while the position score is 5.*

Alignments are given a score based on the quality of the alignment along a path through the matrix.

The extreme version of the initial scoring scheme for amino acids used by Needleman and Wunsch (1970) assigned a positive score of +1 to identical residues and ignored (scored as 0) any mismatches. The total score for the alignment was then determined by summing the scores across aligned positions. The greatest score for a nucleotide sequence of 10 bases is 10 for an identical alignment, and the lowest score is 0 for complete lack of matches. As the sequence length grows, so does the potential score of the alignment. Such a stringent scoring scheme can be useful for finding nearly identical sequences. However, it ignores, by setting to zero, the biologically common existence of mismatches between residues and the presence of insertions and deletions. The extreme scoring scheme, therefore, misses more distantly related gene products.

Local Alignments   Local alignment methods changed the focus of the alignment from finding the best score for a full-length alignment to finding the best alignment score for the longest stretch of residues. Local algorithms determine if there is a fraction of the

---

Highlight 2.3

Global alignments are useful for determining the identity of a gene or sequence. Two sequences are typically considered to have the same identity when >90% of the residue pairs in the alignment are identical.

---

query sequence that can be optimally aligned between query and target sequence (Smith and Waterman, 1981; Wilbur and Lipman, 1983; Altschul *et al.*, 1997). They are not required as with the Needleman and Wunsch approach to align from sequence end to sequence end.

The main changes to the alignment scoring scheme to create a local alignment method in dynamic programming were to (1) require a negative score for mismatches, (2) set any negative alignment score to zero, and (3) terminate any alignment whose score is zero and start over (Fig. 2.6). The alignment is started in the highest scoring matrix and continues until the alignment is terminated. Fine tuning a local alignment algorithm is done by changing the values assigned to matches and mismatches. The optimal alignment is the longest stretch of aligned residues with the highest alignment score.

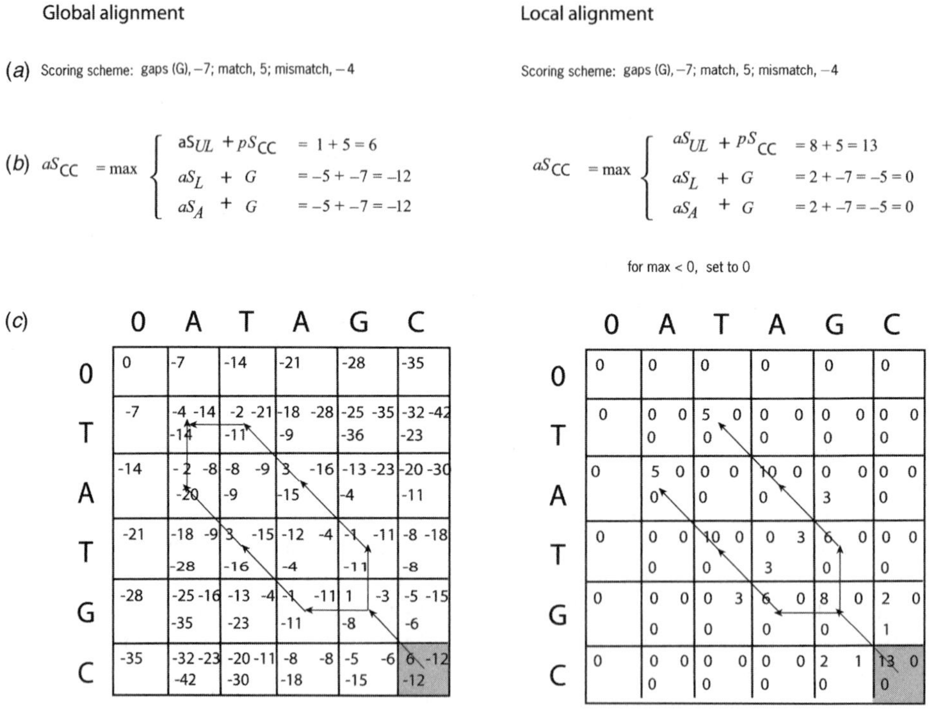

Global alignment

Local alignment

(*a*) Scoring scheme: gaps (G), −7; match, 5; mismatch, − 4

Scoring scheme: gaps (G), −7; match, 5; mismatch, − 4

(*b*) $aS_{CC} = \max \begin{cases} aS_{UL} + pS_{CC} & = 1 + 5 = 6 \\ aS_L + G & = -5 + -7 = -12 \\ aS_A + G & = -5 + -7 = -12 \end{cases}$

$aS_{CC} = \max \begin{cases} aS_{UL} + pS_{CC} & = 8 + 5 = 13 \\ aS_L + G & = 2 + -7 = -5 = 0 \\ aS_A + G & = 2 + -7 = -5 = 0 \end{cases}$

for max < 0, set to 0

(*c*)

Figure 2.6 *(a) Scoring schemes for global (left) and local (right) alignment methods are shown. (b) The difference between scoring methods can be seen in the scoring equations that determine alignment scores in the matrix. The arrows in the populated matrices indicate the path for the optimal alignment traced backwards. (c) The scores from the upper left, above and left cells are written from top left to right and down in each cell. Local alignments begin in the cell with the highest score and substitute a zero for any negative score. Global alignment begin in the lowest right hand cell and negative scores are retained. In this example, the lowest right hand cell is also the highest scoring cell. This is not always the case. Note that there are two equally well-scoring alignments in both matrices.*

2.3.2.2 Gaps The above discussion of global and local alignment methods does not accurately take into account scoring insertions and deletions between sequences that appear as gaps in sequence alignments. Gaps are accounted for by adding a residue location, 0, in the axes of the matrix and by the scoring penalty. The added gap column and row establishes gap penalties for the end of the two sequences, increasing the penalty when the sequences begin to align further away from the ends. The gap score establishes the degree of penalty, a negative score, for having to adjust the alignment by moving down or across the matrix for the highest scoring pair. Thus, the benefit of inserting a gap must be greater than the penalty to the score (Needleman and Wunsch, 1970). When fine tuning global and local alignment methods, it is the relationship between match, mismatch, and gap penalties that primarily determine whether a full-length or partial alignment is retrieved (Vingron and Waterman, 1994).

Detail 2.3

A review of the effect of varying gap, match, and mismatch values on the alignment of random sequences found that use of high mismatch and gap penalties greater than a match score will find local alignments; when the penalty for a mismatch is greater than twice the score for a match, the gap penalty becomes the decisive parameter in the alignment; and for a mismatch penalty less than twice the score of a gap and a wide range of gap penalties (Vingron and Waterman, 1994).

In general, global alignment methods are useful when comparing two sequences to find a measure of similarity or identity over the length of the sequences. Local alignment methods are useful for comparing sequences that have regions of similarity that are only a fraction of the sequence lengths, where one sequence may be a subset of the other or where regions of similarity overlap. Thus, local alignments are better for finding regions of similarity that may arise from conserved domains. These short regions may be missed by global alignments which are tuned to find the best score for the overall length even at the expense of an otherwise high scoring subregion (Mount, 2001; Smith and Waterman 1981; Smith *et al.*, 1981).

2.3.2.3 Searching Sequence Databases

Global Alignments When the sequences are placed within a database, additional issues arise for aligning and scoring sequences: How do we compare alignment results without severely slowing down the search time? The global alignment method is relatively computer intensive. Each pairwise alignment creates a matrix of size $n \times m$, where n is the length of the shorter sequence and m the longer. This formulation originally minimally required the equivalent number of steps ($n \times m$) for the alignment to be computed as well as for the movement of results in and out of memory. This results in a total of $(n \times m)^2$ to be calculated. Although global alignment methods have been improved, they are not typically used to search databases over the Web. Instead, when global alignment algorithms are used remotely, it is likely to compare two sequences provided by the user (Bray *et al.*, 2003; Cochrane *et al.*, 2005). Global alignment tools may be installed and run on local machines when alignments of a large number of sequences are desired.

TABLE 2.2   Local Sequence Alignment Algorithms

Program	Web Site	References
FASTA	http://www.ebi.ac.uk/fasta33/; ftp.virginia.edu	Pearson and Miller (1992); Pearson (1996)
BLAST	http://www.ncbi.nlm.nih.gov/ blast/	Altschul *et al.* (1990)
LALIGN	http://fasta.bioch.virginia.edu/ fasta_www2/ fasta_www.cgi?rm = lalign	Huang and Miller (1991)
Psi-Blast	http://www.ncbi.nlm.nih.gov/ blast/	Altschul *et al.* (1997)
Mega-Blast	http://www.ncbi.nlm.nih.gov/ blast/	Zhang *et al.* (2000)

Local Alignments   Local alignment methods are typically faster than global alignments and often implemented to search remote databases (Table 2.2). A key factor in generating fast search algorithms is to know when to initiate and terminate an alignment with a sequence in the target database. In 1983, Wilbur and Lipman developed the word-based method that is used in many of the popular database search tools. The method is built on the premise that any two sequences with a significant relationship will have at least one word—series of residues—in common. Based on this premise, a word index is generated for sequences within the database. The query sequence is also broken into words that are used to search the index. When matches—now referred to as hits—are made, a sequence alignment with the full target sequence is initiated.

BLAST (basic local alignment search tool) is probably the most well-known local alignment search algorithm. In BLAST, the word index matches identical and neighboring words to the query string. For the word match to be considered a hit, its aligned score must be above a cutoff threshold or be discarded (Altschul *et al.*, 1990, 1997). Once a hit is obtained, the sequence alignment is extended in both directions (Fig. 2.7). More of the query sequence is added to the ends of the sequence fragment until its score can no longer be improved by extension as long as the score stays above the dropoff value X.

Because the local alignment method seeks to determine the maximum alignment in a region of the sequence, multiple regions with high identity or similarity may be found interspersed between regions of low-scoring alignments. As hits are made, the BLAST algorithm determines if the hit is significant in relation to the other hits obtained. The newest alignment score is compared with the maximal score obtained for a previous sequence alignment. The alignment is kept or discarded based on how close it is to the current maximal alignment score. The alignment score in local alignment without gaps

Single hit   Two hits

Figure 2.7   *Schematic illustration of extended sequences. When a hit is made between a query sequence word and the database word index (left), the query sequence length is extended. The original BLAST algorithm looked for single hits between the query word and target to begin alignment extensions (left). The improved gapped BLAST algorithm only extends sequences with two hits (right). This modification allowed for the implementation of gapped alignment searches with retention of search speed (Altschul et al., 1997).*

is of the aligned segment, not the entire sequence. When multiple alignments are found within a single pairwise sequence alignment, the individual scores contribute to an overall sequence alignment score (Wilbur and Lipman, 1983; Karlin and Altschul, 1990).

Returning Local Alignments When searching a database sequence, multiple sequences may be identified that contain locally optimized alignment scores. Altschul *et al.* (1997) called these sequence segments, high-scoring segment pairs (HSPs). The alignments are ordered by their alignment scores and expect values (e-values) (Fig. 2.8). The alignment scores are based on the local alignments or summed local alignments of segments in the two sequences. Not all alignments are returned and presented to the user. An alignment may be discarded as random if its alignment score is low in relation to the overall probability of getting the same alignment score by chance within the database. In BLAST, this is determined by the user-selected e-value at the NCBI search interface, which we discuss further below.

2.3.3 Biological Databases

The number of publicly available databases is increasing rapidly. In addition to sequence databases, databases have been created for organisms, images, RNA expression profiles, signaling and metabolic pathways, and so forth. The question to answer as one selects a database is "What type of data does the database contain?" Databases like NCBI have a well-developed glossary and list of database content. Databases with less-developed user interfaces or help guides require additional research to determine what kind of data is included. The journal *Nucleic Acids Research* (NAR) dedicates the first issue of every year to articles and reviews of database updates and releases. This provides a

```
                                                          Score    E
Sequences producing significant alignments:              (bits)  Value

gi|7485306|pir||T01798  hypothetical protein A_TM021B04.8 - ...   1528   0.0
gi|15240464|ref|NP_198075.1|   expressed protein [Arabidopsis...  1407   0.0
gi|30421188|gb|AAP31312.1|   ABI3-interacting protein 2; CnAI...   108   5e-22
gi|47497800|dbj|BAD19898.1|   putative ABI3-interacting prote...    98   7e-19
gi|47497799|dbj|BAD19897.1|   putative ABI3-interacting prote...    98   1e-18
gi|47497801|dbj|BAD19899.1|   putative ABI3-interacting prote...    97   1e-18
gi|18403383|ref|NP_566709.1|   hydroxyproline-rich glycoprote...    88   8e-16
gi|30695446|ref|NP_850923.1|   expressed protein [Arabidopsis...    86   3e-15
gi|22530976|gb|AAM96992.1|   putative protein [Arabidopsis th...    86   6e-15
gi|33146509|dbj|BAC79626.1|   putative ABI3-interacting prote...    84   2e-14
gi|16974572|gb|AAL31182.1|   AT4g14900/d13490c [Arabidopsis t...    81   1e-13
gi|15222466|ref|NP_174463.1|   expressed protein [Arabidopsis...    81   1e-13
gi|18414336|ref|NP_567447.1|   hydroxyproline-rich glycoprote...    80   2e-13
gi|15237325|ref|NP_197136.1|   expressed protein [Arabidopsis...    75   7e-12
gi|17432946|sp|Q9FDW0|FRI_ARATH  FRIGIDA protein >gi|1080117...     72   6e-11
gi|31558911|gb|AAP49807.1|   FRIGIDA [Arabidopsis thaliana] >...    72   6e-11
gi|31558918|gb|AAP49810.1|   FRIGIDA [Arabidopsis thaliana]         72   6e-11
gi|34394454|dbj|BAC83628.1|   putative hydroxyproline-rich gl...    70   3e-10
gi|34394453|dbj|BAC83627.1|   putative hydroxyproline-rich gl...    70   3e-10
gi|8777369|dbj|BAA96959.1|   unnamed protein product [Arabido...    60   2e-07
gi|7488123|pir||C71412  probable hydroxyproline-rich glycopr...     59   4e-07
gi|34906200|ref|NP_914447.1|   OJ1174_D05.18 [Oryza sativa (j...    52   6e-05
gi|15240463|ref|NP_198074.1|   protein transport protein-rela...    49   8e-04
gi|19922172|ref|NP_610875.1|   CG4840-PA [Drosophila melanoga...    36   5.3    L
```

Figure 2.8 *Results of a BLAST search, arranged by alignment score with corresponding e-values.*

helpful resource for finding databases tailored to one's research interests. These articles include details on database design, sources and methods used to obtain data, and associated computational tools.

The type of information found in molecular sequence databases is essentially a function of three factors: the method by which the data was produced (laboratory experiments, computational analysis); where the data was obtained (direct submission or retrieved from literature on other databases); and its molecular type (DNA, RNA, protein).

Information provided by researchers or extracted from the literature is considered primary data (Baxevanis and Ouellette, 1998) (Fig. 2.9). Data produced by applying computational tools to primary data is secondary data. This is done to discover and predict additional information such as putative phosphorylation sites, transmembrane domains, or identity. A majority of the biological databases contain information based on both experimentation and results of computational analysis; in other words, both primary and secondary data, respectively.

Biological databases that contain primary nucleotide sequence data are GenBank, European Molecular Biology Laboratory (EMBL), and DNA Data Bank of Japan (DDJB). Primary protein sequences are entered to SwissProt and Protein Information Resource (PIR). Databases with secondary data include protein domain and metabolic databases that use the primary sequence data to generate new information. Increasingly, Web sites are developed that aggregate information from multiple sources. NCBI contains sequence data retrieved from SwissProt as well as GenBank, both primary databases. It also contains databases with secondary data, such as RefSeq and Conserved Domain Database (CDD).

The data within primary databases is as reliable as the data submitted, and this depends primarily on the methods used to produce it. Regardless of who obtains the sequence data, genomic sequencing or basic research labs, nucleic and amino acid sequencing results are subject to errors. If the data is the result of a single-pass sequencing of genomic or cloned DNA, it is likely to contain errors. Much of the sequence data obtained from high-throughput labs has yet to be characterized and or validated. NCBI records contain fields that indicate both the method by which sequences were obtained (Express Sequence Tag (EST), Sequence Tagged Site (STS)) as well as their biological nature (genomic, chromosomal, mRNA) (Table 2.3). The types of molecular data that databases contain

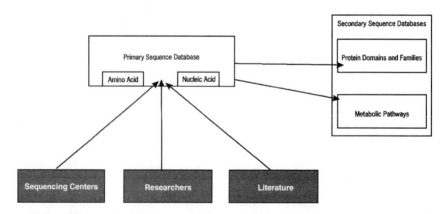

Figure 2.9 *Flowchart of sequence data from labs and literature to primary sequence database and subsequent secondary databases.*

TABLE 2.3 Examples of Molecular Sequence Types in NCBI Records

	Type	Description	Example in Title
Genome	Sequence tagged site (STS)	A unique segment of DNA that occurs only once in a genome and marks a particular location. Can be generated from genomic DNA or cDNA.	"*Homo sapiens* chromosome 1 map 1p35, *sequence tagged site*"
	Draft sequences (phases 0–3)	Pieces of a genome that are compiled from a DNA or cDNA library. They are usually large collections of contigs and are in the process of being ordered and catalogued.	"*Pan troglodytes* clone rp43-26h17, WORKING *DRAFT SEQUENCE*, 46 unordered pieces"
	Genome	The complete genome of an organism.	"*Corynebacterium diphtheriae* NCTC 13129, *complete genome*"
Chromosome	Locus	A known location on a chromosome for a particular gene or collection of genes that codes for a specific function.	"*Takifugu rubripes* isotocin and vasotocin *locus*"
	Contig	A contiguous segment of a chromosome made by joining overlapping clones or sequences.	"*Homo sapiens chromosome* 22 genomic *contig*"
	Chromosome	The whole sequence of a single chromosome.	"*Homo sapiens* chromosome 22, *complete sequence*"
Gene	Domain	A discrete portion of a protein assumed to fold independently of the rest of the protein and which possesses its own function.	"V-like *domain* binding molecule"
	Complete CDS	A complete coding sequence (CDS) for a protein.	"*Mus musculus* gene for G protein–coupled receptor TGR7, *complete cds*"
	Gene	Whole gene sequence for a protein or proteins.	"*Homo sapiens* GNAS1 *gene*"
mRNA Note: all cDNA sequences are represented as mRNA	Expressed sequence tag (EST)	A partial sequence of cDNA in mRNA form from either the beginning (5′) or the end (3′) of a protein or gene sequence.	"*Karenia brevis EST Library* (L99-05) *Karenia brevis* cDNA, mRNA sequence"
	Complementary DNA sequence (cDNA)	A cDNA sequence in mRNA form. A cDNA sequence is originally coded from mRNA, so it is a genomic DNA sequence but without the introns and noncoding regions.	"*Mus musculus cDNA sequence* BC034834 (BC034834), mRNA"
	Complete CDS	A complete mRNA sequence for a protein coding region.	"*Rattus norvegicus* SCIRP10-related protein (SCIRP10) mRNA, *complete cds*"

Source: NCBI.

differ: genomic, chromosomal, mRNA, protein, and so forth. Performing a nucleotide-based sequence alignment search of a database containing this many types of nucleic acids and their corresponding sequence records is likely to produce multiple alignments.

NCBI hosts a series of databases that contain one or more data type. It has aggregated multiple resources that make it possible for biologists to obtain basic molecular information about a gene or protein of interest fairly quickly. At the same time that the aggregation of data makes it very useful, it also means that the absence of a well-constructed search can result in a substantial if not overwhelming number of retrieved data records. Although our conversation will focus mostly on the use of NCBI, one is encouraged to examine other data and computational resources. As one's needs and question become more refined, the use of more tailored resources may improve one's search results while decreasing overall search time. These interests may be tailored by organism, molecular type (i.e., full mRNA sequences, genomic, EST, partial sequences, etc.) or degree of characterization (putative gene, putative or known cellular function).

2.3.3.1 *BLAST at NCBI*

We will now look at the implementation of Basic Local Alignment Search Tool, a.k.a. BLAST, at NCBI (Altschul *et al.*, 1990). BLAST is one of the most well-known sequence search algorithms. It has been integrated into many freely available Web-based databases where the code has been licensed so that it can be used to search these databases for similar sequences. The NCBI BLAST page (http://ncbi.nlm.nih.gov/BLAST) provides a list of databases that can be searched within NCBI.

The original BLAST algorithm was developed for ungapped, local sequence alignments. It has since been modified to optimize the chances of finding a broader range of sequence alignments with database searches. The additional alignments include nucleic to protein acid sequences (tBLASTn), protein to protein sequence (BLASTp) and protein to nucleic (tBLASTx). The modified BLAST algorithms are called subroutines. This first set of subroutines take into account the nucleic and amino acid codes and automate the translation of one code to another (Table 2.4).

Minimum Requirements   To run a BLAST search, the user enters the query sequence to be compared against the database and selects a database to be searched (Fig. 2.10). A query can be entered to the search field by using an NCBI accession number or the sequence.

TABLE 2.4 BLAST Subroutines

Program	Database	Query	Comments
blastp	Protein	Protein	Uses substitution matrix for finding distant relationships; SEG filtering available
blastn	Nucleotide	Nucleotide	Tuned for very-high-scoring matches, not distant relationships
blastx	Protein	Nucleotide (translated)	Useful for analysis of new DNA sequences and ESTs
tblastn	Nucleotide (translated)	Protein	Useful for finding unannotated coding regions in database sequences
tblastx	Nucleotide (translated)	Nucleotide (translated)	Useful for EST analysis

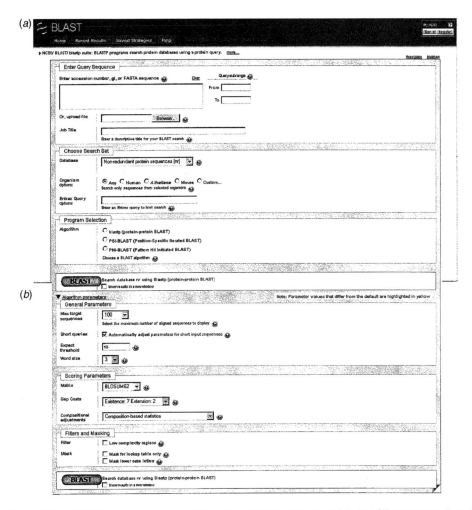

Figure 2.10 *BLASTp interface accessed through protein blast link on NCBI BLAST home page (http:// www.ncbi.nlm.nih.gov/BLAST). (a) By entering a query sequence (Enter Query Sequence), selecting a database (Choose Search Set), and search algorithm (Program Selection), a BLAST search can be performed. (b) The "Algorithm parameters" link opens a set of menus to modify parameter values. The parameters have been assigned three categories: General Parameters, Scoring Parameters, and Filters and Masks.*

Sequences stored in .txt files can be uploaded or copied and pasted into the BLAST search field. When using an NCBI accession number, the search engine locates the record and uses the sequence within the record. For example, the NCBI record NM_169061 contains the sequence of the *Drosophila melanogaster* IP3 receptor. Rather than copying the sequence to a separate file, we can type in the NCBI accession number or the gene identifier number (GI|24644260). The assignment of unique identifiers makes it possible for sequences to be retrieved at the BLAST search interface.

The default database for protein BLAST searches is the *nr* database. This is the historical name of the database which used to be curated to remove redundant sequences. *nr* stood for nonredundant. The *nr* database is also available for nucleotide searches. It contains records of nucleotide sequences from multiple databases: GenBank, EMBL, DDBJ, and protein data bank (PDB). It does not contain low pass sequences such as expressed sequence tags (EST), genomic survey sequences (GSS) or high throughput genomic sequences (HTGS). These latter sequence types can be found in separate databases or "genomic + transcripts" databases for specific organisms.

Multiple high-scoring sequence alignments may be retrieved when blasting due to:

- Partial sequences: records of overlapping sequences are not removed from the database.
- Multiple sources: the nucleotide sequences may be genomic, mRNA, or expressed sequence tags and each has a separate record.
- Different species: sequence records from multiple species.
- Redundant sequences: sequences that differ by one nucleotide are treated as different sequences in NCBI.

One approach to reducing the number of retrieved sequences is to limit the search space by choosing the database most specific to the purpose of a search. Limits similar to those discussed for keywords are also available for BLAST searches at NCBI.

Each BLAST search can be fine tuned by varying the BLAST program parameters (Table 2.5). The tunable parameters include selecting a substitution scoring matrix, word size, gap penalties, filters, sequence composition, and acceptable expect score. We introduce the BLAST algorithm parameters that can be varied to optimize sequence similarity searches for one's purposes (Fig. 2.10).

Word Size BLAST aligns two sequences relatively quickly by first searching for matches between words generated from the query sequence and those in the database index of target sequences. The word size sets the number of nucleotides to be used to search the database. The word index matches identical and neighboring words to the

TABLE 2.5 Adjustable Parameters for BLAST Search at NCBI

Blast Field	Purpose
Expect	Sets the e-value limit that you are willing to accept in the found matches. The higher the e-value, the more distantly related sequences may be found and the more likely a random sequence match could obtain similar alignment scores.
Word size	Sets the minimum initial residue length for beginning an alignment: align three residues, four, and so forth. More hits are likely to be obtained with smaller word sizes.
Matrix	Scoring matrices (PAM or BLOSUM) used to determine alignment score between two sequences.
Gap costs	Penalty value for the presence and length of gaps in the sequence alignment. The set value contributes to the overall alignment score.
Compositional adjustments	Used in protein sequence alignments to take into account proteins whose amino acid composition has a significantly different amino acid frequency that what is assumed in the substitution matrices (PAM and BLOSUM).
Filters	Filters are employed to mask areas of low complexity in nucleotide sequences, e.g. GC or AT rich regions. This reduces obtaining high scoring matches due to low complexity regions.

query string. Each successful alignment of the query word to a target sequence is known as a hit. For the alignment to be considered a hit, its aligned score must be above a cutoff threshold or be discarded (Altschul *et al.*, 1997).

Scoring: Substitution Scoring Matrices   The scoring process is similar to the process mentioned previously for pairwise local alignments. Residue pairs within the aligned segments are scored based on identical, mismatched, or gapped pairs. The scoring schemes in BLASTp, protein-protein sequence search, use substitution matrices: PAM and BLOSUM (Dayhoff, 1978; Henikoff and Henikoff, 1992). The substitution matrices are reference tables of residue match scores assigned based on the probability of a substitution occurring for any given residue (Fig. 2.11).

The matrices tune the search algorithm to find either highly conserved (high similarity) or very divergent sequences (lower similarity) (Fig. 2.12). PAM (percent accept mutation) matrices predict a frequency of substitutions based on assumed evolutionary distances (Dayhoff *et al.*, 1978; Schwartz and Dayhoff, 1978). A PAM 1 matrix predicts the frequency of a given substitution when only 1 mutation in 100 sites occurs; that is, 1% of amino acids change (Dayhoff *et al.*, 1978). The value assigned to the substitution is based on its evolutionary probability. For example, a conservative nucleotide or amino acid substitution, a hydrophobic amino acid for another, will occur more often within highly related sequences than a nonconservative substitution, hydrophilic for hydrophobic

BLOSUM62

	C	S	T	P	A	G	N	D	E	Q	H	R	K	M	I	L	V	F	Y	W	
C	9																				C
S	-1	4																			S
T	-1	1	5																		T
P	-3	-1	-1	7																	P
A	0	1	0	-1	4																A
G	-3	0	-2	-2	0	6															G
N	-3	1	0	-2	-2	0	6														N
D	-3	0	-1	-1	-2	-1	1	6													D
E	-4	0	-1	-1	-1	-2	0	2	5												E
Q	-3	0	-1	-1	-1	-2	0	0	2	5											Q
H	-3	-1	-2	-2	-2	-2	1	-1	0	0	8										H
R	-3	-1	-1	-2	-1	-2	0	-2	0	1	0	5									R
K	-3	0	-1	-1	-1	-2	0	-1	1	1	-1	2	5								K
M	-1	-1	-1	-2	-1	-3	-2	-3	-2	0	-2	-1	-1	5							M
I	-1	-2	-1	-3	-1	-4	-3	-3	-3	-3	-3	-3	-3	1	4						I
L	-1	-2	-1	-3	-1	-4	-3	-4	-3	-2	-3	-2	-2	2	2	4					L
V	-1	-2	0	-2	0	-3	-3	-3	-2	-2	-3	-3	-2	1	3	1	4				V
F	-2	-2	-2	-4	-2	-3	-3	-3	-3	-3	-1	-3	-3	0	0	0	-1	6			F
Y	-2	-2	-2	-3	-2	-3	-2	-3	-2	-1	2	-2	-2	-1	-1	-1	-1	3	7		Y
W	-2	-3	-2	-4	-3	-2	-4	-4	-3	-2	-2	-3	-3	-1	-3	-2	-3	1	2	11	W
	C	S	T	P	A	G	N	D	E	Q	H	R	K	M	I	L	V	F	Y	W	

Figure 2.11 *BLOSUM 62 matrix. The observed frequency of amino acid pairs is calculated from sequences in the BLOCKS database, which are grouped by percent identity. The observed frequency is divided by the so-called frequency of the pair occurring by chance. The log of this quotient, observed frequency over chance frequency, is then taken and entered as the score for the aligned pair.*

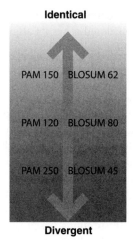

Figure 2.12   *Diagram indicating sensitivity of PAM and BLOSUM matrices for divergent sequences.*

(Dayhoff *et al.*, 1978; Altschul, 1991; States *et al.*, 1991; Henikoff and Henikoff, 1992). The frequencies of the substitution was established in the creation of PAM 1. The frequency for a given amino acid substitution (i.e., alanine for threonine) in the PAM 250 matrix is extrapolated from the initial PAM 1 by assuming the same initial frequency occurring independently 250 times, (i.e., PAM 1 multiplied against itself 250 times). The PAM 250 creates better scoring alignments for distantly related proteins (14–27%) similarity than lower PAM matrices (George *et al.*, 1990).

The block substitution matrices, or BLOSUM, are also based on the frequency of amino acid substitutions between sequences. However, instead of predicting the amount of substitutions between sequences based on evolutionary distances, the frequencies were calculated from the number of substitutions found empirically within highly conserved sequence stretches called blocks. For example, the number and type of substitutions in sequences that were 62% identical were used to generate the BLOSUM 62 matrix (Henikoff and Henikoff, 1992). The values given to specific residue pairs in BLOSUM are based on the observed frequency of that substitution in sequences with a specific percentage of identity. BLOSUM 62 has been shown experimentally to be good for finding weak protein similarities (Henikoff and Henikoff, 1992).

Gap Costs   Related and not necessarily identical sequences, when aligned, have insertions and deletions that appear as gaps in the alignment. If an alignment with few gaps is desired (i.e., highly similar sequences or regions), the aligned sequences should be penalized in their score for having gaps. If greater divergence is desired, gaps are acceptable and sequences should be penalized less. The gap penalties in BLAST allow the user to set the negative value used for the presence and length of gaps in the alignment. As mentioned earlier, it is the fine tuning of gap penalties, match and mismatch scores that affect both length and quality of the alignments returned. When using local alignment methods, a high gap penalty for the presence and length of gaps can result in short, highly conserved alignments. As the penalty for the presence of gaps is decreased, longer extensions are expected.

Gapped alignments at NCBI are the default method (Altschul *et al.*, 1997). The severity of the penalty is set by two factors: that a gap exists, "Existence," and the length of the gap,

"Extension." Because a single mutational event may cause the insertion or deletion of more than one residue, the presence of a gap is frequently ascribed more significance than the length of the gap (Vingron and Waterman, 1994). Gap penalties are selected from the drop down menu for the "Gap Costs" (Fig. 2.10).

Filters   DNA sequences contain regions of low complexity. These are areas in which the base composition is dominated by one or two nucleotides (e.g., GC-rich regions or Alu sequences in DNA). The presence of low-complexity regions can result in sequence alignments having high similarity scores because of the alignment of these regions between DNA sequences (Altschul *et al.*, 1994). The filters available with BLAST identify regions with the query sequence that are characterized as low complexity and instead substitute Xs for amino acid and Ns for nucleic acid residues in order to bypass high scores from aligning these regions (Wootten and Fedheren, 1996).

Expect   The expect parameter in NCBI's BLAST sets statistical significance threshold for sequence alignments. The e-value (expect value) is a statistical measure for each alignment that determines the probability of obtaining an equivalent alignment score—bit or S score—if a random sequence of the same length were used with the same BLAST parameter settings (Karlin and Altschul, 1990). A higher e-value says that there is a high probability that a random sequence of the same length would obtain the same or better alignment score. The lower the e-value, the less likely another sequence with a better score will be found in the database, and therefore the score is more significant. By setting an e-value at the BLAST interface, one limits the displayed alignments to those with the same e-value or lower.

The e-value is the calculated likelihood of aligning a random sequence to any of the thousands to hundreds of thousands of sequences in the database (Karlin and Altschul, 1990). It answers the question for two proteins idealized as a random ordering of independently selected amino acids: "What is the likelihood of obtaining an alignment score, S, within the database?" In the BLAST algorithm, the expect value can be used as a significance threshold.

2.3.3.2 *Displaying Results*   The number of records retrieved depends on the database used and the type of search done. By performing a BLAST search of the nucleotide or protein sequence database, sequences containing "significant" alignments based on one's parameter settings are returned. The format options on the NCBI BLAST search pages set the display options for one's results. The default settings provide a graphical view of the highest scoring alignments, followed by a list of record names with links to the corresponding alignments and target sequence record. The number of alignments shown below the graphical display depends on the value one chose in format options for the number of returns and expect values (Fig. 2.13). NCBI also provide links to other databases that it hosts when additional information concerning the alignment or retrieved target sequence is available. For nucleotide searches, the links may be to UniGene or LocusLink. Proteins may have links to the NCBI domain or protein structure databases.

The NCBI visualization tool plots the number of residues of the query sequence as a bar scale of length (Fig. 2.13A). Aligned sequences are shown below the bar. These segments range from red bars, indicating alignment scores of >200 that extend the full length of the query sequence, to black slashes dotted over the length of the query, indicating the absence of alignment between stretches of high scoring pairs. The longer alignments are often splice variants of nucleotide queries, protein isoforms, or redundant sequence records

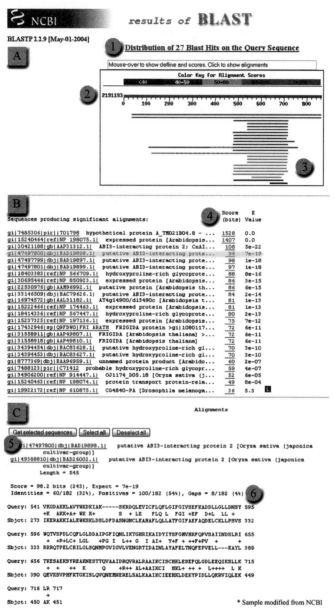

Figure 2.13 *Results of a BLAST query. (A) Graphical results: (1) The number of regions from the target database that aligned with the query sequence. Multiple regions can be identified for a single hit sequence. (2) The query sequence with span of base pairs indicated. If a unique identifier is entered for the query sequence, the gene accession number is displayed to the left. (3) The graphical representation of the hit regions compared with the query sequence. Regions are color-coded according to the bit score of the region. Regions that are connected with a hatched line represent a single hit sequence with multiple regions of alignment. Detached regions on the same line represent unrelated hits. Clicking on a region will take one to the sequence alignment. (B) List of records producing hits. (4) List of hit identifiers and scores in order of e-value. Clicking the identifier takes one to the full sequence record. Clicking a score takes one to the sequence alignment. (C) Alignment information for hit highlighted in (B). (5) Multiple identifiers indicate redundant records for the target sequence and are counted as one sequence hit. (6) The score and characteristics of the query/target sequence alignment.*

(i.e., GenBank, RefSeq). The shorter, lower similarity regions may arise from common sequence motifs such as phosphorylation sites or putative kinase domains that are conserved across proteins and not specific to its overall identity. When such motifs are found, NCBI provides a link to its in-house domain database CDD as well as external records in protein family or domain databases.

List of Alignments   The alignments are listed with their sequence ID, the database source, accession number for the sequence record, followed by the organism and assigned name (Fig. 2.13B). Each sequence record is shown with the alignment score (bit score) and e-value. The bit score measures the quality of the alignment between the query and retrieved sequence, only taking into account the region or regions that are aligned within the pair. The lowest e-value possible is 0 and is typically associated with nearly identical sequences (e.g., splice variants, isoforms, and redundant records). Less significant e-values (higher numbers) are obtained with shorter stretches of alignment and as the overall quality of the alignment decreases. Any alignment with an e-value greater than 0.02 has a 20% chance of obtaining the same alignment score by searching with a random sequence. These alignments should be examined thoroughly before accepting it as something statistically or biologically significant.

NCBI Search Statistics   NCBI provides a set of statistics regarding one's search. The number of hits, attempted and successful gapped extensions is of particular interest for evaluating the success of a sequence alignment search and BLAST parameter settings. The number of hits as well as the extension threshold reflects the success of the word size on generating hits in the database. If very few hits are generated, a smaller word size may be chosen to increase the hit number. By generating more hits, the number of attempted extensions will also increase. Gap penalties and substitution matrices also affect overall extension and quality of alignments.

Detail 2.4
    Reading score and characteristics of the query/target sequence alignment.

    Identities   Number of base pairs that match out of the total and add to score. Identities
        are indicated by the same letter between the query and target sequences.
    Positives   Number of identities plus the number of acceptable substitutions as
        determined by the substitution matrix, both of which add to the score. Acceptable
        substitutions are indicated by plus signs (+) on the sequence alignment.
    Gaps   Number of single residue gaps in the query and target sequences. Gaps are
        indicated by a dash (−) on the sequence alignment.

2.4   CONCLUSION

The text here provided a brief introduction to the basis of pairwise sequence alignment and searches for statistically significant alignments within biological databases. We attempt here to provide both background information on the development of sequence alignment methods and instruction for the commonly used search tool, BLAST.

Algorithms tailored to search for similar sequences within databases are typically developed with rules designed to speed up the search process. The rule in BLAST is to wait until two hits are obtained within a given sequence before performing a full alignment. The developers have shown that this rule retains sensitivity while increasing search speed. The trade-off is that such tools may miss biologically significant alignments. Once target sequences are identified using search tools, it is important to run a series of confirming tests: (1) use the target sequence as a query sequence; (2) randomize query sequence and see if the same sequence is obtained; (3) run a global or local alignment algorithm independent of the database. The latter process allows the compute time to be dedicated to performing a full search for the optimal alignment.

For the scale of sequence alignment and database searches discussed in this chapter, very little compute resources are required. Bioinformatics resources are frequently made available via the Web as exemplified by the NCBI resource. Additionally, the National Science Foundation and other funding agencies have invested in research resources—data storage, compute time, and software—that are publicly available. As such, typically initial research can be performed with a Web browser and ethernet connection, 10 Megabits (Mbits) per second. These same resource providers have workshops and tutorials available for members of the research community to learn about and make use of these tools.

More in-depth discussions of alignment methods including hidden Markov models and advanced scoring schemes can be found in books dedicated to sequence analysis and in molecular sequence–focused bioinformatics texts (Baxevanis and Ouellette, 1998; Durbin et al., 1998; Mount, 2001). As well, the methods of sequence alignment and database searches have become sufficiently standard that protocols are now available for the lab.

BIBLIOGRAPHY

Altschul SF (1991). Amino acid substitution matrices from an information theoretic perspective. *Journal of Molecular Biology* 219:555–565.

Altschul SF (1993). A protein alignment scoring system sensitive at all evolutionary distances. *Journal of Molecular Evolution* 36:290–300.

Altschul SF, Gish W (1996). Local alignment statistics. *Methods in Enzymology* 266:460–480.

Altschul SF, Gish W, Miller W, *et al.* (1990). Basic local alignment search tool. *Journal of Molecular Biology* 215(3):403–410.

Altschul SF, Boguski MS, Gish W, Wootton JC (1994). Issues in searching molecular sequence databases. *Nature Genetics* 6(2):119–129.

Altschul SF, Madden TL, Schaffer AA, *et al.* (1997). Gapped BLAST and PSI-BLAST: a new generation of protein database search programs. *Nucleic Acids Research* 25:3389–3402.

Baxevanis AD, Oullette BFF, eds. (1998). *Bioinformatics: A Practical Guide to the Analysis of Genes and Proteins.* In: *Methods of Biochemical Analysis*, Vol 39. New York: Wiley-Interscience; 370p.

Bedell JA, Korf I, Gish W (2000). MaskerAid: a performance enhancement to RepeatMasker. *Bioinformatics* 16(11):1040–1041.

Bishop MJ (1999). *Genetic Databases.* San Diego: Academic Press; 295p.

Bray N, Dubchak I, Pachter L (2003). AVID: a global alignment program. *Genome Research* 13: 97–102.

Cochrane G, Aldebert P, Althorpe N, *et al.* (2006). EMBL Nucleotide Sequence Database: developments in 2005. *Nucleic Acids Research* 34(Database issue):D10–D15.

Dayhoff MO, Schwartz RM, Orcutt BC (1978). A model of evolutionary change in proteins. In: Dayhoff MO, ed. *Atlas of Protein Sequence and Structure, Vol. 5, Suppl. 3.* Washington, D.C.: National Biomedical Research Foundation; pp. 345–352.

Durbin R, Eddy SR, Krogh A, Mitchison G (1998). *Biological Sequence Analysis: Probabilistic Models of Protein and Nucleic Acids.* Cambridge: Cambridge University Press; 356p.

George DG, Barker WC, Hunt LT (1990). Mutation data matrix and its uses. *Methods in Enzymology* 183:333–351.

Gruber TM, Eisen JA, Gish K, *et al.* (1998). The phylogenetic relationships of Chlorobium tepidum and Chloroflexus aurantiacus based upon their RecA sequences. *Federation of European Microbiological Societies Microbiology Letters* 162(1):53–60.

Henikoff S, Henikoff JG (1992). Amino acid substitution matrices from protein blocks. *Proceedings of the National Academy of Sciences USA* 89:10915–10919.

Huang X, Miller W (1991). A time-efficient, linear-space local similarity algorithm. *Advances in Applied Mathematics* 12:337–357.

Kan Z, Rouchka EC, Gish WR, *et al.* (2001). Gene structure prediction and alternative splicing analysis using genomically aligned ESTs. *Genome Research* 11(5):889–900.

Karlin S, Altschul SF (1990). Methods for assessing the statistical significance of molecular sequence features by using general scoring schemes. *Proceedings of the National Academy of Sciences USA* 87(6):2264–2268.

Karlin S, Altschul SF (1993). Applications and statistics for multiple high-scoring segments in molecular sequences. *Proceedings of the National Academy of Sciences USA* 90(12): 5873–5877.

Karlin S, Bucher P, Brendel V, *et al.* (1991). Statistical methods and insights for protein and DNA sequences. *Annual Review of Biophysics and Biophysical Chemistry* 20:175–203.

Ma B, Tromp J, Li M (2002). PatternHunter: faster and more sensitive homology search. *Bioinformatics* 18(3):440–445.

Marth GT, Korf I, Yandell MD, *et al.* (1999). A general approach to single-nucleotide polymorphism discovery. *Nature Genetics* 23(4):452–456.

Mount D (2001). *Bioinformatics: Sequence and Genome Analysis.* Cold Spring Harbor: Cold Spring Harbor Laboratory Press; 564p.

Needleman SB, Wunsch CD (1970). A general method applicable to the search for similarities in the amino acid sequence of two proteins. *Journal of Molecular Biology* 48(3):443–453.

Pearson WR (1995). Comparison of methods for searching protein sequence databases. *Protein Science* 4:1145–1160.

Pearson WR (1996). Effective protein sequence comparison. *Methods in Enzymology* 266:227–258.

Pearson WR (2000). Flexible sequence similarity searching with the FASTA3 program package. *Methods in Molecular Biology* 132:185–219.

Schwartz RM, Dayhoff MO (1978). Matrices for detecting distant relationships. In: Dayhoff MO, ed. *Atlas of Protein Sequence and Structure, Vol. 5, Suppl. 3.* Washington, D.C.: National Biomedical Research Foundation; pp. 353–358.

Smith TF, Waterman MS (1981). Identification of common molecular subsequences. *Journal of Molecular Biology* 147:195–197.

Smith TF, Waterman MS, Fitch WM (1981). Comparative biosequence metrics. *Journal of Molecular Evolution* 18:38–46.

States DJ, Botstein D (1991). Molecular sequence accuracy and the analysis of protein coding regions. *Proceedings of the National Academy of Sciences USA* 88:5518–5522.

States DJ, Gish W, Altschul SF (1991). Improved sensitivity of nucleic acid database searches using application-specific scoring matrices. *Methods* 3:66–70.

Vingron M, Waterman MS (1994). Sequence alignment and penalty choices: review of concepts, case studies and implications. *Journal of Molecular Biology* 235:1–12.

Westhead DR, Parish JH, Twyman RM (2002). Bioinformatics. In: Hames BD, ed. *Instant Notes Series*. New York: BIOS Scientific Publishers Limited; 304p.

Wilbur WJ, Lipman DJ (1983). Rapid similarity searches of nucleic acid and protein data banks. *Proceedings of the National Academy of Sciences USA* 80:726–730.

Wootton JC, Federhen S (1996). Analysis of compositionally biased regions in sequence databases. *Methods in Enzymology* 266:554–71.

Zhang Z, Schwartz S, Wagner L, *et al.* (2000). A greedy algorithm for aligning DNA sequences. *Journal of Computational Biology* 7(1-2):203–214.

Chapter 3

Family-Domain Databases

Once a protein sequence is obtained, there are many questions that can be asked. What is the protein's overall identity? What putative functions does it have? What biological motifs are present? A number of databases have been tailored to help answer these questions. In this chapter, we will look at the types of information that can be obtained regarding protein putative functions via family and domain databases. Protein identity is typically determined by performing global dynamic programming sequence alignments with related sequences. The methods for predicting structural and functional features of proteins are designed to exploit characteristic sequence patterns and amino acid frequency and properties. Different computational tools are needed to determine possible functional domains based on primary sequence data.

Family and domain databases are used to address the question, "What domains are contained within this sequence? or what family does this protein belong to?" Although, some family and domain databases were developed with the intent to annotate genomic sequences, basic researchers are also interested in using these tools to better characterize their proteins of interest. To answer the questions of what families the sequence belongs to or what domains does it contain, we must first define what we mean by *families* and *domains*. In the following sections on protein families and domains, we discuss how the biological patterns are defined and modeled in secondary databases such that new information is obtained.

3.1 DEFINITIONS: FAMILY AND DOMAIN

3.1.1 Biological Definitions

It is helpful to begin our discussion of family and domain databases by defining key words that are used to describe the data they contain. The biologically significant terms used in the family-domain databases are *family*, *domain*, and *motif*. Each identifies biological characteristics that are conserved across a set of proteins and associated with a biological functionality.

> A *family* of proteins was originally defined by Dayhoff *et al.* (1978) as a group of sequences with more than 50% identity when aligned with similar function. Families are often also characterized by the presence of one or more domains with high sequence similarity.
>
> *Domains*, traditionally known as structurally independently folding units, are conserved functional units that may contain one or more motifs.
>
> *Motifs* are conserved across proteins at the level of sequence or structure or both. They include both short stretches of fixed residue length that act as sites for post-translational modifications, phosphorylation, and longer sequences that form secondary structures for protein-DNA, protein-ion, or protein-lipid interactions. The terms *consensus sequence*, *pattern*, and *signature* are generally used to refer to the sequence or structural traits that define a motif, domain, or family (Table 3.1).

Protein families are made up of proteins related to one another by sequence similarity, domain composition, or structure. These include proteins found across species (orthologues) or within the same species (paralogs). Once families are identified experimentally, familial descriptors can be created for identifying and classifying future family members. The family description is typically derived from multiple sequence alignments (MSAs) that enable us to define traits that encompass all member sequences. Family descriptors have been based on sequence identity (e.g., >50% identical); common domains (e.g., kinase catalytic domain, calcium binding motifs, etc.); structure (e.g., bundle of seven transmembrane helices); or a combination of these characteristics.

Domains represent discrete stretches within the protein, unlike protein families, which are commonly defined over the length of the sequence (Fig. 3.1). Domains are functional, structural units within proteins. The structural definition of a domain is an independent folding unit (e.g., alpha helices) (Fig. 3.1). These units are conserved at the level of sequence and structure. Domain databases formalize the descriptions of domains by defining combinations of short regions of highly conserved amino acids within a domain; domain length descriptions that take into account all amino acids; or the domain's structural features. Domain traits are delineated in the same fashion as protein families: multiple sequence alignments are created, and a domain description is developed that accounts for its pattern of amino acids.

An implied characteristic of all motifs is that it is a pattern retained in homologous proteins that is the basis for homologous functions (Blundell *et al.*, 1983). Thus, motifs such as SH2 domains, calcium binding domains, and plecktrin repeats are found in multiple

TABLE 3.1 Term

General Identifiers	Biological Object	Defined By
Pattern, signature, consensus	Family	Sequence, structure
	Domain	Sequence, structure
	Motif	Sequence

(*a*) Members of the Cadherin Family

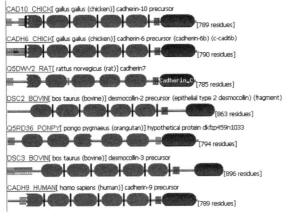

Sequence Titles	CAD1_CHICK	EPITHELIAL-CADHERIN PRECURSOR (E-CADHERIN) (LIVER CELL ADHESION MOLECULE) (L-CAM) - Gallus gallus
	CAD1_HUMAN	EPITHELIAL-CADHERIN PRECURSOR (E-CADHERIN) (UVOMORULIN) (CAM 120/80) - Homo sapiens (Human).
	CAD1_MOUSE	EPITHELIAL-CADHERIN PRECURSOR (E-CADHERIN) (UVOMORULIN) (ARC-1) - Mus musculus (Mouse).
	CAD1_XENLA	EPITHELIAL-CADHERIN PRECURSOR (E-CADHERIN) (UVOMORULIN) - Xenopus laevis (African clawed frog).
	CAD2_CHICK	NEURAL-CADHERIN PRECURSOR (N-CADHERIN) - Gallus gallus (Chicken).
	CAD2_HUMAN	NEURAL-CADHERIN PRECURSOR (N-CADHERIN) - Homo sapiens (Human).
	CAD2_MOUSE	NEURAL-CADHERIN PRECURSOR (N-CADHERIN) - Mus musculus (Mouse).
	CAD3_HUMAN	PLACENTAL-CADHERIN PRECURSOR (P-CADHERIN) - Homo sapiens (Human).
	CAD3_MOUSE	PLACENTAL-CADHERIN PRECURSOR (P-CADHERIN) - Mus musculus (Mouse).
	CAD4_CHICK	RETINAL-CADHERIN PRECURSOR (R-CADHERIN) (R-CAD) - Gallus gallus (Chicken).
	CAD4_MOUSE	RETINAL-CADHERIN PRECURSOR (R-CADHERIN) (R-CAD) - Mus musculus (Mouse).
	CAD5_HUMAN	VASCULAR ENDOTHELIAL-CADHERIN PRECURSOR (VE-CADHERIN) (CADHERIN-5) (7B4 ANTIGEN) (CD144 ANTIGEN)
	CAD5_MOUSE	VASCULAR ENDOTHELIAL-CADHERIN PRECURSOR (VE-CADHERIN) (CADHERIN-5) - Mus musculus (Mouse).
	CAD6_HUMAN	CADHERIN-6 PRECURSOR (KIDNEY-CADHERIN) (K-CADHERIN) - Homo sapiens (Human).
	CAD6_MOUSE	CADHERIN-6 PRECURSOR (KIDNEY-CADHERIN) (K-CADHERIN) - Mus musculus (Mouse).

(*b*) Sample architecture

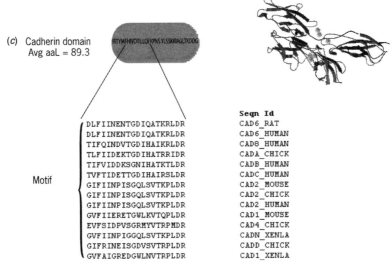

CAD10_CHICK[gallus gallus (chicken)] cadherin-10 precursor
[789 residues]

CADH6_CHICK[gallus gallus (chicken)] cadherin-6 precursor (cadherin-6b) (c-cad6b)
[790 residues]

Q5DWV2_RAT[rattus norvegicus (rat)] cadherin7
Cadherin_C [785 residues]

DSC2_BOVIN[bos taurus (bovine)] desmocollin-2 precursor (epithelial type 2 desmocollin) (fragment)
[863 residues]

Q5RD36_PONPY[pongo pygmaeus (orangutan)] hypothetical protein dkfzp459n1033
[794 residues]

DSC3_BOVIN[bos taurus (bovine)] desmocollin-3 precursor
[896 residues]

CADH9_HUMAN[homo sapiens (human)] cadherin-9 precursor
[789 residues]

(*c*) Cadherin domain
Avg aaL = 89.3

IRTYMFHIVDTILLQEKPNSYLSSKKIAGLTKDDG

Motif	**Seqn Id**
DLFIINENTGDIQATKRLDR	CAD6_RAT
DLFIINENTGDIQATKRLDR	CAD6_HUMAN
TIFQINDVTGDIHAIKRLDR	CAD8_HUMAN
TLFIIDEKTGDIHATRRIDR	CADA_CHICK
TIFVIDDKSGNIHATKTLDR	CADB_HUMAN
TVFTIDETTGDIHAIRSLDR	CADC_HUMAN
GIFIINPISGQLSVTKPLDR	CAD2_MOUSE
GIFIINPISGQLSVTKPLDR	CAD2_CHICK
GIFIINPISGQLSVTKPLDR	CAD2_HUMAN
GVFIIERETGWLKVTQPLDR	CAD1_MOUSE
EVFSIDPVSGRMYVTRPMDR	CAD4_CHICK
GVFIINPIGGQLSVTKPLDR	CADN_XENLA
GIFRINEISGDVSVTRPLDR	CADD_CHICK
GVFAIGREDGWLNVTRPLDR	CAD1_XENLA

Figure 3.1 *Examples of protein families, domain, and motif. The protein family is cadherin. Shown are titles of protein records for members with SwissProt (release 34.0) naming convention taken from PRINTS. Sample architectures obtained from Pfam containing the cadherin domain. Icon for cadherin domain shown next to structural image from PDB record 1edh of the cadherin domain (Nagar et al., 1996). Also shown is a structural depiction of the cadherin domain. PRINTS motif for cadherin domain, only one of seven motifs shown.*

proteins. Because of the reuse of motifs and domains, similarities can be found within sequences that are otherwise unrelated evolutionarily. Motif searches often involve a highly conserved and relatively short stretch of amino acids compared against a fairly large database. A pairwise alignment search (e.g., BLAST or FASTA) is vulnerable to losing hits as a result of the alignment score of a small stretch of amino acids being too small relative to statistical significance (Tatusov *et al.*, 1994). As well, if a hit is returned, a number of false positives are statistically likely to be found because of the size (i.e., number of residues) in the database. Because proteins can be functionally related and have very little conserved sequence similarity (e.g., globins), methods are needed to distinguish between similarities due to random variation and those of common origin or function (Lesk and Clothia, 1980; Bork and Gibson, 1996).

The family-domain databases provide the following benefits:

Increased sensitivity in database searches: Sensitivity refers to the ability of the search algorithm to detect all true matches. Search methods for identifying motifs or domains in a sequence typically use information derived from multiple sequence alignments. This enables the search algorithm to identify biologically and statistically significant similarities that might be missed with a single local alignment search with a "representative" sequence (Henikoff, 1996).

Increased specificity for identifying functionally similar proteins: Specificity refers to the ability to detect only related proteins. Functionality occurs predominately at the level of structures within domains (Blundell *et al.*, 1983; Sweet, 1986). Basing searches for similarity on domain representations increases the specificity of the search for functionally related proteins.

Classification of protein sequences: The domain databases create an indexed access to protein sequences, in which sequences are classified according to family assignments.

3.2   FAMILY-DOMAIN DATABASES

Family and domain databases were developed to help catalogue and automate the detection of protein domains and family members. Both types of databases identify families of proteins; however, they differ in the formalized definition of family. For the protein family databases, the focus is on the characterization of the protein sequence as an entire unit, in which protein domains are a part. For domain databases, domain families consist of a set of proteins that are related based on the presence of the domain in the sequence. Both involve the characterization of domains, however the change in emphasis influences the classification methods used. We will refer to the set of databases defining family and domains as the family-domain databases and focus our discussion on the characterization of domains.

We focus on methods used to develop a few classic and more popular databases. These databases include PROSITE, BLOCKS, PRINTS, Pfam, and Simple Modular Architecture Tool (SMART) (Table 3.2). It is helpful to know the biological focus of the different databases. The majority of these databases focus on classifying families, subfamilies, and domains. Subfamilies are those proteins that share greater similarity with one another than with other members of the family and yet are still members of the family.

TABLE 3.2 Database References

Name	Web Address	Description	Reference
PROSITE	http://www.expasy.ch/prosite	Groups of proteins of similar biochemical function on basis of amino acid patterns	Hulo *et al.* (2006)
PRINTS	http://www.bioinf.man.ac.uk/dbbrowser/PRINTS/	Protein fingerprints or sets of unweighted sequence motifs from aligned sequence families	Attwood *et al.* (2003)
BLOCKS	http://blocks.fhcrc.org/	Ungapped blocks in families defined by the ProSite catalogue	Henikoff *et al.* (1999); Henikoff *et al.* (2000)
Pfam	http://www.sanger.ac.uk/Pfam	Profiles derived from alignment of protein families, each one composed of similar sequence and analyzed by hidden Markov models	Finn *et al.* (2006)
SMART	http://smart.emblheidelberg.de/	Genetically mobile domains	Letunic *et al.* (2004)
InterPro	http://www.ebi.ac.uk/interpro	Integrated resource of protein domains and functional sites: combination of Pfam, PRINTS, ProSite, and current SwissProt/TrEMBL sequence	Mulder *et al.* (2005)
PIR	http://www.nbrf.georgetown.edu/pirwww/index.shtml	Family and superfamily classification based on sequence alignment	Wu *et al.* (2003)

Myoglobins and hemoglobins are two subfamilies of the larger globin family. The majority of myoglobins will show greater similarity of structure and sequence with each other than they will with a hemoglobin.

PROSITE classifies both family and subfamilies and lists the subfamily representation within the family documentation. Pfam attempts to classify large families and makes no distinctions between subfamilies. PRINTS develops individual representations for each subfamily using BLOCKS. BLOCKS characterizes highly conserved motifs. The primary focus of SMART is on the development of domains such that domain families are determined rather than protein families. Each database develops a model of the family that is a characterization of the motifs and domains within the proteins (Table 3.3). These models are then used to further classify and identify additional sequences. A family descriptor is useful when it is sensitive enough to detect divergent members yet specific enough to detect only family members.

The family descriptors have been based on the domain composition and order of discrete conserved regions, or by a matrix representation of the full-length sequence. Identifying multiple domains within a sequence increases the sensitivity of the algorithm for proper classification of divergent sequences (Sigrist *et al.*, 2002). When motifs are found in the context of multiple, ordered motifs, the likelihood of detecting an accurate

TABLE 3.3   Approaches to Defining Families and Domains

Database	Approach
PROSITE	Families or domains defined by conserved regions
PRINTS	Families defined by multiple conserved regions
BLOCKS	Domains defined by highly conserved regions
Pfam	Families or domains defined by entire domains
SMART	Families defined by multiple entire domains

fingerprint of a protein family increases despite the possible presence of low-scoring individual motifs (Scordis *et al.*, 1999; Wright *et al.*, 1999; Attwood *et al.*, 2003, 2005). The assertion that the ability to identify multiple domains of signaling proteins improves sensitivity for such proteins was the premise for building SMART (Shultz *et al.*, 1998; Letunic *et al.*, 2006).

3.3   CREATING DOMAIN REPRESENTATIONS

Multiple strategies have been employed with family-domain databases to achieve the goal of accurate inclusion and exclusion of sequences as family members. The strategies are distinguished by their choices within the multistaged process of creating the database. The steps are (1) selection of initial sequences, (2) method of multiple sequence alignment, (3) type of representation of family, and (4) method of comparison or matching. Based on the choices made at each step, different results may be obtained from the databases. We spend the next pages reviewing each step and the methods employed by all or some of the databases in their process of classifying protein family-domains.

3.3.1   Initial Sequences

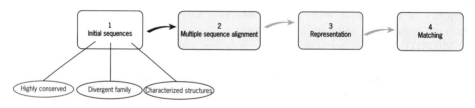

Step 1 of creating family-domain database.

An "initial set of sequences" is selected to establish the defining characteristics of members of the protein family, domain, or motif. As shown in Table 3.4, the selection criteria for initial sequences, also known as seed sequences, have varied from one database to another. PROSITE develops a pattern for proteins that have been identified in the literature as functionally or structurally similar (Sigrist *et al.*, 2002). BLOCKS selects sequences based on their degrees of similarity (Henikoff and Henikoff, 1994). SMART sequences were selected from proteins with known 3D structures (Schultz *et al.*, 1998; Letunic *et al.*, 2006).

TABLE 3.4   Initial Sequences

Database	Includes	Selection Method
PROSITE	Sequence and structure	Manual
PRINTS	Sequence	Automated
BLOCKS	Sequence	Manual and automated
Pfam	Sequence	Semiautomated
SMART	Sequence and structure	Automated

In order to identify distantly related members, the initial sequences used to construct the family representation ideally contain at least one representative of divergent sequences. By selecting sequences that include divergent sequences, we set the groundwork for creating the multiple sequence alignments (MSAs) and representations that can detect distantly related sequences. Additional strategies for increasing representation of distantly related sequences within the initial set of alignments include (1) limiting the number of identical sequences used within the alignment, (2) using other inclusion criteria other than sequence similarity (i.e., structure), and (3) increasing the scoring weight of more divergent sequences within the MSA.

3.3.2   Multiple Sequence Alignment

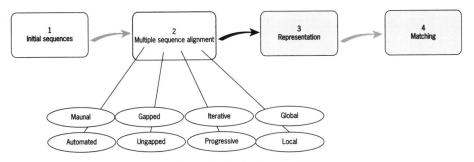

Step 2 of creating family-domain database.

Multiple sequence alignment methods provide a more sensitive diagnostic tool than pairwise alignments. Multiple methods for MSA exist (Table 3.5). They may be manual or automated, iterative or progressive, and obtain gapped or ungapped, local or global alignments (Bairoch, 1992; Henikoff and Henikoff, 1992; Thompson *et al.*, 1994; Lassmann and Sonnhammer, 2002). The family-domain databases typically employ automated or semiautomated methods, which use automated alignment tools to create an initial MSA and follow it with a manual examination of the alignment quality. CLUSTAL W, a global progressive alignment method, is the most commonly implemented method in the databases discussed in this chapter (Thompson *et al.*, 1994).

Progressive alignment methods begin with a series of pairwise alignments to create a phylogenetic tree that is then used as a guide tree for the MSA (Fig. 3.2). There are three different approaches to constructing the phylogenetic tree: maximum parsimony, distance matrix, and maximum likelihood. The tree nodes and branches closely duplicate the sequence data in terms of differences in residues between sequences. CLUSTAL W uses a

TABLE 3.5   Progressive Multiple Alignment Methods

Program	Alignment Type	Representation
Clustal W[a]	Global	Generalized profile
POA[b]	Local	Graph
T-Coffee[c]	Local	Words

[a]Thompson *et al.*, 1994.
[b]Grasso and Lee, 2004.
[c]Notredame *et al.*, 2000.

distance matrix method that first determines the global similarity between sequences using the general substitution matrices (BLOSUM 80). The neighbor-joining method is then used for tree construction (Saitou and Nei, 1987). Closely related sequences are aligned first at either the root or branches of the tree, and then more distantly related sequences are added (Feng and Doolittle, 1987; Thompson *et al.*, 1994).

3.3.3   Representations: Domains and Families

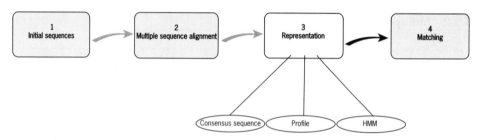

Step 3 of creating family-domain database.

Once the MSA is obtained from the initial sequences, the protein family and domain traits are derived and represented as one of the following: alignment blocks, consensus sequence, expression pattern, position-specific scoring matrix, profile or hidden Markov model (Gribskov *et al.*, 1987; Baldi *et al.*, 1994; Krogh *et al.*, 1994; Tatusov *et al.*, 1994; Eddy *et al.*, 1995; Henikoff, 1996) (Table 3.6). These different representations can be classified as creating representations for short stretches of highly conserved residues or for all residues of the characterized protein or protein region within the MSA.

Block, consensus sequence, and expression patterns are representations for discrete regions of highly conserved amino acids, not the entire domain (Fig. 3.3). The term *block* and the corresponding BLOCKS database were developed by the creators of BLOSUM weight matrices (Henikoff and Henikoff, 1992). Blocks are short, ungapped stretches of amino acids that are conserved across sequences with different percentages of identity (80%, 62%, etc.). A consensus sequence as defined by Gribskov *et al.* (1987) is a composite representation of the amino acid usage in the motif, derived from the amino acid frequencies and PAM score at given positions within the protein sequence. Expression patterns are defined by PROSITE as stretches of strongly, but not absolutely, conserved residues separated by characteristic spacing.

(a)   Amino acid sequences

Seq. A   YNIIRRHKHKHRNAIQVGT
Seq. B   YHLIRPHQHNHRNAMEVGT
Seq. C   YNLISPYKLKLRNAIEVGT
Seq. D   YNLLRPYKLKLRNAIEVGI
Seq. E   CNLIRPYKHKLRNAIEVGT
Seq. F   YNLLRPYKLKLRSAIEVGT

(b)   Distance matrix

	Seq. A	Seq. B	Seq. C	Seq. D	Seq. E	Seq. F
Seq. A	0	7	7	8	6	8
Seq. B		0	8	9	7	9
Seq. C			0	3	3	3
Seq. D				0	4	2
Seq. E					0	4
Seq. F						0

(c)   Guide tree

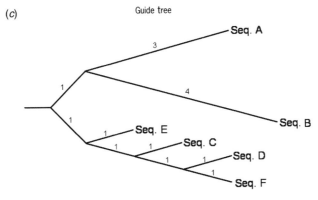

Figure 3.2 *Example of a phylogenetic tree created using a distance matrix. (a) Six amino acid sequences of the same length. (b) The distance matrix shows the number of non-identical residues between each pair of sequences. (c) The pairwise distances are used to construct the phylogenetic tree. Sequences and distance matrix were created in EvolSeq 1.2 (Weisstein and Jungck, 2006), available online through the Biological ESTEEM Project at http://bioquest.org/esteem.*

TABLE 3.6   Database Method of Representation

Database	Method of Representation
BLOCKS	Alignment of conserved region in multiple sequence (blocks); motif HMM
PRINTS	Multiple blocks; motif HMM
PROSITE	Regular expression pattern; profiles
Pfam	Profile HMM
SMART	Entire profile HMM; consensus

In contrast, position-specific scoring matrices (PSSM), profiles, and HMMs develop models in the form of matrices that characterize all residue positions in the length of the protein family or domain within the MSA. By describing all residues, the matrices create a representation encompassing the nonconserved regions as well as the conserved. The result of having representations of the entire domains is increased sensitivity for divergent family members (Gribskov *et al.*, 1987; Krogh *et al.*, 1994; Gribskov and Veretnik, 1996; Eddy, 1998).

The relative distribution or the frequency of residue usage in a position within the motif can be different than what is expected for amino acids when position-specific information is not taken into account (i.e., based on PAM or BLOSUM). The PSSMs are derived from the frequency of amino acid usage at each position of the aligned sequences. One axis of the matrix is the length of the aligned sequences. The opposite coordinate contains the 20 amino acid residues and gaps. A score is assigned based on whether and how often (frequency) the amino acid appears within the aligned sequences and the random probability that the amino acid appears. The probability for any amino acid is 1/20. The completed matrix provides a score for any amino acid at each position of the aligned sequences. The matrix is a specific scoring matrix for the family rather than a general substitution matrix for all proteins.

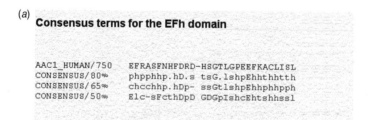

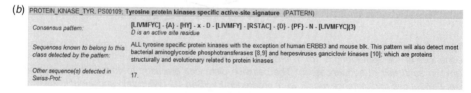

Figure 3.3 *Illustration of consensus patterns. (a) The consensus sequence for the EFh domain is taken from SMART. (b) The PROSITE database develops expression patterns that reflect the consensus sequence of the aligned family. The pattern is for tyrosine protein kinases.*

Highlight 3.1

The position specific scoring matrix (PSSM), first proposed by Gribskov as profile analysis (Highlight Fig. 3.1.1), introduced a method to take into account the frequency of amino acid usage within aligned sequences. The aligned amino acid sequences are used to generate a matrix of average frequencies of amino acid usage at each position, $W(b, p)$. Gribskov then combined the average frequency with the Dayhoff frequency counts for all proteins, $Y(b, p)$ to produce the matrix of position specific scores, $M(p, a)$. Thus, the biologically occurring mutation frequency of amino acids as described by Dayhoff is accounted for together with the averaged frequency of amino acids in the aligned sequences. The position specific score for a given amino acid, a, at position, p, is formally described as $M(p, a) = \sum_{b=1}^{20} W(p, b) \times Y(a, b)$.

POS	PROBE	CONSENSUS	A	C	D	E	F	G	H	I	K	L	M	N	P	Q	R	S	T	V	W	Y	+/-
1	EGVL	V	3	-2	3	4	0	4	-1	3	-1	4	4	1	1	1	-2	1	2	6	-6	-2	9
2	LLSP	L	2	-2	-2	-1	3	0	-1	3	-1	6	5	-1	3	0	-1	3	1	4	1	-1	9
3	VVVV	V	2	2	-2	-2	2	2	-3	11	-2	8	6	-2	1	-2	-2	0	2	15	-9	-1	9
4	KEAT	A	6	-2	5	6	-5	4	1	0	5	-2	0	3	3	3	1	3	6	0	-6	-4	9
5	APLP	P	6	-1	0	1	-2	2	0	1	0	2	2	0	8	2	0	2	2	3	-5	-4	9
6	GGGG	G	7	1	7	5	-6	15	-1	-3	0	-4	-3	4	3	2	-3	6	4	2	-11	-7	9
7	SSQE	D	4	-1	7	7	-6	7	2	-2	2	-3	-2	4	3	6	1	6	2	-1	-6	-5	9
8	SSTP	S	4	4	2	2	-4	4	-1	0	2	-3	-2	2	7	0	1	10	6	0	-2	-4	9
9	VLVA	V	5	0	-1	-1	3	1	-2	7	-2	7	6	-1	1	-1	-3	0	2	10	-5	-1	9
10	KRRS	R	0	-1	1	1	-5	0	2	-2	8	-3	1	3	3	3	10	5	1	-2	7	-5	9
11	MLII	I	0	-2	-3	-2	7	-3	-3	11	-1	11	10	-2	-2	-1	-2	-2	1	9	-3	1	9
12	SSTS	S	4	6	2	2	-3	5	-1	0	2	-3	-2	3	4	-1	1	12	6	0	0	-4	9
13	CCCC	C	3	15	-5	-5	-1	2	-1	3	-5	-8	-6	-3	1	-6	-3	7	3	3	-13	10	9
14	KSQR	K	1	-2	3	3	-6	1	3	-2	7	-3	0	3	3	5	7	4	1	-2	2	-5	9
15	AAGS	A	10	3	4	3	-5	8	-1	-1	1	-2	-1	3	4	1	-2	7	4	2	-6	-4	9
16	TSDS	S	4	3	5	4	-5	6	0	0	2	-3	-2	4	3	1	1	9	6	0	-3	-4	9
17	GGSQ	G	5	1	6	5	-6	9	1	-2	1	-3	-2	4	3	4	0	6	3	0	-6	-6	9
18	YFLS	P	-1	2	-4	-3	9	-3	0	4	-3	6	3	-1	-3	-3	-3	1	-1	2	7	7	9
19	TTRL	T	1	-2	0	1	0	0	0	2	2	2	3	1	1	1	3	1	7	2	1	-2	9
20	PF.L	P	-2	-3	-6	-4	10	-4	-1	6	-4	9	6	-3	-4	-4	-3	-2	-1	3	7	8	4
21	SS.D	S	3	2	5	4	-4	5	0	-1	2	-3	-2	4	3	1	1	8	2	-1	-2	-3	4
22	S..S	S	2	3	1	1	-2	3	-1	0	1	-2	-1	2	2	0	1	8	2	0	1	-2	4
23	...G	G	2	0	2	1	-2	4	0	0	0	-1	-1	1	1	1	1	2	1	1	-3	-2	4
24	...D	D	1	-1	4	3	-2	2	1	0	1	-1	-1	2	1	2	0	1	1	0	-3	-1	4
25	...G	G	2	0	2	1	-2	4	0	0	0	-1	-1	1	1	1	-1	2	1	1	-3	-2	4
26	.AGN	A	6	0	4	3	-4	6	1	-1	1	-2	-1	5	2	2	-1	3	3	1	-5	-3	4
27	YNYT	Y	0	5	0	-1	5	-1	2	1	-1	0	-1	4	-3	-2	-2	0	3	0	3	6	4
28	EDDY	D	2	-2	9	8	-3	3	4	-1	1	-3	-2	5	-1	4	-1	1	1	-1	-6	0	9
29	LMAL	L	3	-5	-3	-1	6	-1	-2	6	-1	10	10	-2	0	0	-2	-1	0	6	-1	0	9
30	YNAW	N	4	1	3	2	0	2	3	-1	1	-1	-1	8	0	1	-1	2	1	-1	-1	2	9
.	.	.																					
.	.	.																					
48	SGNS	S	4	3	5	3	-4	7	0	-2	2	-4	-3	6	3	1	0	10	3	0	-2	-4	9
49	SSNY	S	2	5	2	1	1	2	1	0	1	-2	-2	5	1	-1	0	8	1	-1	3	1	9

Highlight Figure 3.1.1 Sample profile, position specific scoring matrix. Four probe sequences of the immunoglobulin variable-region domain are aligned vertically to the left (PROBE). The 49 residue positions (POS) are listed to the left and a derived CONSENSUS sequence to the right. The PROFILE has 21 columns, one for each amino acid and the rightmost column of the profile gives the penalty for insertion/deletion (+ / −). Positions 31–47 of the profile were omitted from the figure for clarity. This figure and modified legend was reproduced with permission from Gribskov et al., 1987. PNAS 84, 4355–4358.

Terms for Matrix of Position Specific Scores Defined

$$W(b, p) = \frac{n(b, p)}{N_R}$$

The average frequency is determined by number of times that the amino acid b appears at position p in the aligned sequences over the number of residues.

$n(b, p)$ = the number of times that the amino acid b appears at position p in the aligned sequences. Position p is on the y-axis and amino acid b is on the x-axis of the matrix.

N_R = the number of residues in the aligned sequences is equivalent to the length of the sequences. The residue positions appear as rows. The profile matrix shows an abbreviated row of 49 residues.

$Y(b, p)$ = Matrix of mutational frequency in proteins such as PAM-Dayhoff (see Chapter 2).

Weighting particular sequence contributions within the MSA is done to ensure that sequences that are underrepresented in number (i.e., due to fewer orthologues than paralogs) are equally represented. Different weighting schemes are used to establish the contribution of individual sequences and their residues to the scoring matrix. These schemes are designed to increase sensitivity of the PSSM to divergent member sequences (Henikoff and Henikoff, 1992; Vingron and Sibbald, 1993; Luthy *et al.*, 1994; Thompson *et al.*, 1994). The original PSSM and profiles were modified to take into account the rates of substitutions that occur in all proteins and on evolutionary scales. Incorporating the general substitution scores (Dayhoff PAM scores) increased the sensitivity of the specific scoring matrix to divergent members of the family (Gribskov and Veretnik, 1996; Henikoff and Henikoff, 1996). Evaluation of these schemes has led to the consensus that it does not matter significantly which scheme is used but rather that one be used (Henikoff and Henikoff, 1996).

Increasingly, HMMs are used to model families and search databases. A hidden Markov model (HMM) is a probabilistic model of the family derived by aligning an initial set of sequences and determining the probability of any amino acid being matched, deleted, or inserted in any given position within the sequences (Eddy, 1998). This is known as the profile HMM, which is used by PROSITE, Pfam, and SMART. HMMs are considered to be one of the more inclusive and accurate methods for identifying protein family members. The accuracy of the HMM, similar to all representations, is dependent on the sequences used to build the representation and the parameter values selected for cutoff and inclusion of sequences as matches to the model.

3.3.4 Matching: Finding New Members

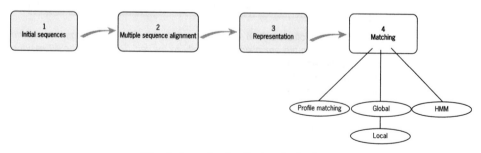

Step 4 of creating family-domain database.

TABLE 3.7 Primary Protein Sequence Databases

Database	Content
SwissProt	All known protein sequences
SwissProt/	Known protein sequences and
TrEMBL[a]	translated sequences from EMBL
UniPROT	Sequence-based protein families
BAliBASE	Structurally aligned MSAs
Pretab	Structurally aligned MSAs

[a]SwissProt/TrEMBLE is searched by Pfam, SMART, and PROSITE (Lutenic *et al.*, 2004).

Once sequences are chosen, alignments made, and representation created, the developers use the family-domain database representations to search for known and unknown family members within sequence databases or test data sets (Edgar, 2004; Thompson *et al.*, 2005) (Table 3.7). The search determines if the family descriptor improves the sensitivity or specificity for finding family members. Already characterized family members serve as a positive control and measure of sensitivity. Previously uncharacterized sequences are evaluated to determine if they are true matches (an indication of increased sensitivity) or false positives (decreased specificity).

3.3.4.1 Scores The match score is the score of an alignment between query and motif or domain. Match scores are generated by moving the matrix along the sequence and determining the best alignment score possible. When the scoring matrices (profiles, HMMs, and PSSMs) are used to match sequences, the alignment may be global or local. A local alignment is a match to part of the scoring matrix, and a global alignment requires the entire representation to be matched.

Because the methods of representation and searches differ from one database to another, the statistical measures of significance also differ slightly. The common characteristics are that a match score is generated between the query sequence and domain representation, created with a variant of dynamic programming. The match score is then used to determine a measure of the probability that it was obtained by chance and a measure of the likelihood of obtaining the score based on the size of the database (e-value). Recall from Chapter 2, the higher the e-value the more likely a match score could be obtained by chance (Karlin and Alschul, 1990, 1993).

3.3.4.2 Cut off The alignment scores between the domain representation and sequences in the database are used to generate a distribution. The score distribution gives some indication of how specific or sensitive the representation is for family members. Ideally, one obtains a predominantly bimodal distribution (Fig. 3.4) with related sequences distributed at one end and unrelated sequences at another. The distribution scores are then used to select typically two threshold values that define the family, known as a gather cutoff and trusted cutoff. The lower of the two is a score that identifies all likely matches to the domain. The higher cutoff score corresponds to the lowest match score for a known member of the family. Despite the attempt to tune cutoff values to highly sensitive, yet family specific settings, the exact value is arbitrary and its success depends on the mode/representation and the distribution of hit scores. Empirically selected cutoff values are likely to produce the best results for achieving highly sensitive and specific matches.

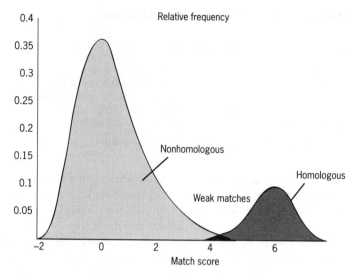

Figure 3.4 *Illustration of distribution of sequences by match score. Those with a high score are considered homologues, while those with a low score are not. Sequences falling between may be weak matches.*

3.3.4.3 e-values The e-value is a statistical measure for each match that determines the probability of obtaining an equivalent score—bit or raw score—if a random sequence of the same length were used to search the database (Karlin and Altschul, 1990). A higher e-value says that there is a high probability that a random sequence of the same length would obtain the same or better alignment score. The lower the e-value, the less likely another sequence with a better score will be found in the database and therefore the score is more significant. The likelihood of a score being found by chance increases proportionally with the size of the database. Match scores for sequences to families and domains that are modeled by PSSMs or HMMs are converted to log odds values that can be used to calculate the e-value.

Proteins may have more than one e-value. One e-value is for the overall sequence and the others for the domains within the sequence. The sequence e-value score in databases such as PRINTS is the sum of all domain scores. Thus, an overall e-value for a sequence becomes higher with each additional domain found. Such a sequence containing multiple domains may have a bad e-value score, i.e. high score or greater than 0.05, and an acceptable gathering cutoff score for the family due to its match score. Conversely, in Pfam, a low domain e-value may be returned to the user because the entire sequence contains multiple domains. It is generally accepted that the presence of multiple domains increases the likelihood that a domain does exist within the sequence.

3.3.5 The Records

The family-domain databases maintain at least two types of records. One is the domain record that contains the initial sequences, MSA, and representation of the family or domain. The second type of record is sequence records of proteins that have been found to be related to the family-domain records. The family representation is used to search against primary protein sequence databases to identify related sequences that were not

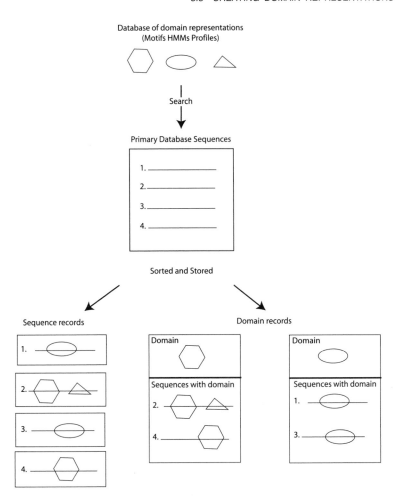

Figure 3.5 *Schematic of domain database searching. Domain database representations are used to search sequence records for matches. The results are stored as part of domain and sequence records.*

included in the initial sequences used to create the family-domain representation (Fig. 3.5). The domain records are subsequently updated with the newly found matches and the individual sequence records of the matches are stored in the database.

3.3.5.1 Family-Domain Assignments: True-False, Positive-Negative
In most of the databases, annotation of the records is an automated process that is curated afterward. The uncurated results of the search may be true or false predictions. The four categories of possible results are true-positive, false-positive, false-negative, or true-negative result. These precalculated results are stored in the database as part of the domain or sequence records.

The category assignments are based on both empirical data—sequences already known to be in the family—and the match score of the alignment. A true-positive hit includes sequences that are known experimentally to have the domain and those sequences that have alignment scores significantly higher than the threshold for matching the domain. False positives are matches to the representation that occur by chance. Manual examination of sequence alignment and biological context determine that the sequence does not contain the domain. Negative results are sequences that fall below the gather thresholds values for the family. False negatives are those sequences that are known to have the domain, presumably from experimental results, and yet were not found by the domain representation and search algorithm.

Highlight 3.2

PROSITE begins with a manual alignment of proteins identified as related through similar sequence, structure, and function. A profile or expression pattern is generated that is used to classify new sequences as containing similar or dissimilar subsequences.

(a) According to PROSITE record PS00109, protein tyrosine kinases have an active site defined by the pattern [LIVMFYC]-x-[HY]-x-D-[LIVMFY]-[RSTAC]-x(2)-N-[LIVMFYC](3). The signature is read as either LIVMFY or C is the first residue, followed by any residue (x), then D, any one of LIVMF or Y, any one of RSTAC followed by any two residues, a single N, and any one of LIVMFYC three times (Highlight Fig. 3.2.1*a*).

  PROSITE refers to patterns capable of finding matches to the family-domain as a signature. The signature is used to search the UniProt/SwissProt database. Based on the number of false positives retrieved it is either deemed an appropriate core signature or expanded in length to decrease the number of false positives identified (PROSITE: http://expasy).

(b) A list of proteins identified from a search of the Swiss-Prot database as having the tyrosine kinase pattern are noted in the PROSITE record (Highlight Fig. 3.2.1*b*).

(c) Domain records are annotated with the results of sequence database searches. The results are organized into lists positive, false positive, potential, and false negative sequences (Highlight Fig. 3.2.1*c*).

(*a*) **NiceSite View of: PS00109**

General information about the entry	
Entry name	PROTEIN_KINASE_TYR
Accession number	PS00109
Entry type	PATTERN
Date	APR-1990 (CREATED); APR-2006 (DATA UPDATE); SEP-2006 (INFO UPDATE)
PROSITE documentation	PDOC00100
Name and characterization of the entry	
Description	Tyrosine protein kinases specific active-site signature.
Pattern	[LIVMFYC]-{A}-[HY]-x-D-[LIVMFY]-[RSTAC]-[D]-{PF}-N-[LIVMFYC](3).

Numerical results

- UniProtKB/Swiss-Prot release number: 50.8, total number of sequence entries in that release: 234112
- Total number of hits in UniProtKB/Swiss-Prot: 435 hits in 435 different sequences
- Number of hits on proteins that are known to belong to the set under consideration: 418 hits in 418 different sequences
- Number of hits on proteins that could potentially belong to the set under consideration: 3 hits in 3 different sequences
- Number of false hits (on unrelated proteins): 14 hits in 14 different sequences
- Number of known missed hits: 9
- Number of partial sequences which belong to the set under consideration, but which are not hit by the pattern or profile because they are partial (fragment) sequences: 7
- Precision (true hits / (true hits + false positives)): 96.76 %
- Recall (true hits / (true hits + false negatives)): 97.89 %

(b)

Cross-references

True positive hits:

```
7LESS_DROME  (P13368),  7LESS_DROVI  (P20806),  AATK_HUMAN  (Q6ZMQ8),
AATK_MOUSE   (Q80YE4),  ABL1_CAEEL   (P03949),  ABL1_HUMAN  (P00519),
ABL1_MOUSE   (P00520),  ABL2_HUMAN   (P42684),  ABL_CALVI   (P11681),
ABL_DROME    (P00522),  ABL_FSVHY    (P10447),  ABL_MLVAB   (P00521),
ACK1_HUMAN   (Q07912),  ACK1_MOUSE   (O54967),  ACK2_BOVIN  (O02742),
ALK_HUMAN    (Q9UM73),  ALK_MOUSE    (P97793),  BMX_HUMAN   (P51813),
BMX_MOUSE    (P97504),  BTKL_DROME   (P08630),  BTK_CHICK   (Q8JH64),
BTK_HUMAN    (Q06187),  BTK_MOUSE    (P35991),  BUD32_YEAST (P53323),
CAD96_DROME  (Q9VBW3),  CSF1R_BRARE  (Q9I8N6),  CSF1R_FELCA (P13369),
CSF1R_HUMAN  (P07333),  CSF1R_MOUSE  (P09581),  CSF1R_RAT   (Q00495),
CSFR1_FUGRU  (P79750),  CSFR2_FUGRU  (Q8UVR8),  CSK_CHICK   (P41239),
CSK_HUMAN    (P41240),  CSK_MOUSE    (P41241),  CSK_RAT     (P32577),
DDR1_HUMAN   (Q08345),  DDR1_MOUSE   (Q03146),  DDR1_PANTR  (Q7YR43),
DDR1_RAT     (Q63474),  DDR2_HUMAN   (Q16832),  DDR2_MOUSE  (Q62371),
EGFR_DROME   (P04412),  EGFR_HUMAN   (P00533),  EGFR_MOUSE  (Q01279),
EGL15_CAEEL  (Q10656),  EPA4A_BRARE  (O13148),  EPA4A_XENLA (Q91845),
EPA4B_XENLA  (Q91694),  EPB1A_XENLA  (Q91571),  EPB1B_XENLA (Q91736),
EPHA1_HUMAN  (P21709),  EPHA1_MOUSE  (Q60750),  EPHA2_HUMAN (P29317),
EPHA2_MOUSE  (Q03145),  EPHA3_BRARE  (O13146),  EPHA3_CHICK (P29318),
EPHA3_HUMAN  (P29320),  EPHA3_MOUSE  (P29319),  EPHA3_RAT   (O08680),
EPHA4_CHICK  (Q07496),  EPHA4_HUMAN  (P54764),  EPHA4_MOUSE (Q03137),
EPHA5_CHICK  (P54755),  EPHA5_HUMAN  (P54756),  EPHA5_MOUSE (Q60629),
EPHA5_RAT    (P54757),  EPHA6_HUMAN  (Q9UF33),  EPHA6_MOUSE (Q62413),
EPHA6_RAT    (P54758),  EPHA7_CHICK  (O42422),  EPHA7_HUMAN (Q15375),
EPHA7_MOUSE  (Q61772),  EPHA7_RAT    (P54759),  EPHA8_HUMAN (P29322),
EPHA8_MOUSE  (O09127),  EPHA8_RAT    (P29321),  EPHB1_CHICK (Q07494),
EPHB1_HUMAN  (P54762),  EPHB1_RAT    (P09759),  EPHB2_CHICK (P28693),
```

(c) **False negative hits (sequences which belong to the set under consideration, but which have not been picked up by the pattern or profile):**

```
BLK_HUMAN    (P51451),  BLK_MOUSE    (P16277),  EPHAA_HUMAN (Q5JZY3),
EPHB2_COTJA  (Q90344),  ERBB3_HUMAN  (P21860),  ERBB3_PONPY (Q5RB22),
FGFR3_BRARE  (Q9I8X3),  VAB1_CAEEL   (O61460),  WSCK_DROME  (P83097)
```

'Potential' hits (partial sequences which belong to the set under consideration, but which are not hit by the pattern or profile because they are partial (fragment) sequences):

```
EGFR_CHICK   (P13387),  EGFR_MACMU   (P55245),  ERBB4_MOUSE (Q61527),
IGF1R_PIG    (Q29000),  INSRR_RAT    (Q64716),  INSR_MACMU  (Q28516),
LCK_RAT      (Q01621)
```

Sequences which could potentially belong to the set under consideration:

```
GCP_METJA    (Q58530),  GCP_METTH    (O27476),  GCVK_HCMVT  (Q68101)
```

False positive hits (sequences which do not belong to the set under consideration):

```
KKA1_ECOLI   (P00551),  KKA1_SALTY   (Q03447),  KKA5_STRFR  (P00555),
KKA8_ECOLI   (P14509),  KKA9_STRRI   (P13250),  MPK12_ARATH (Q8GYQ5),
MPK2_ORYSA   (Q5J4W4),  MSP1_PLAFK   (P04932),  MSP1_PLAFW  (P04933),
PKN6_MYXXA   (P54738),  TITIN_HUMAN  (Q8WZ42),  VATC_THEVO  (Q97CQ2),
Y396_THEAC   (Q03021),  Y444_METJA   (Q57886)
```

Retrieve an alignment of UniProtKB/Swiss-Prot true positive hits:

[Clustal format, color, condensed view] [Clustal format, color] [Clustal format, plain text] [Fasta format]

Highlight Figure 3.2.1

A uniform distribution of sequence scores in the database is one reason for obtaining false positives or negatives. The cutoff value for inclusion is one point within the distribution of sequence scores in the database. When the distribution of scores is not sharp enough to distinguish between homologues and non-homologues, the cutoff threshold may falsely include sequences. Match scores that fall within the "questionable" zone of the distribution matrix are not clearly members or nonmembers. Pfam refers to matches with such scores as potential hits. Potential hits are also created by a local match, only part of the matrix aligned well to the sequence. Thus, potential hits indicate that a domain may be present in the query sequence but the information is insufficient to make a solid classification.

3.3.6 Searching Family-Domain Databases

Most of the family-domain databases support searches for domain records by their data fields: name, ID, or biological process. To find matches between a newly identified sequence and existing families or domains requires comparing the sequence to the family or domain library within the database. In order to analyze a sequence for the presence of already defined domains, we use the sequence as a query. Because each database implements its own variation of domain definition, representation, and search method, different and often complementary predictions for the presence of domains can be produced. It is ideal to query multiple databases to determine if a domain is predicted within one's sequence.

InterPro, a web-based database, was designed to integrate data from complementary databases in order to fully annotate protein families and domains. It obtains data from major family and domain databases including ProDom, Pfam, SMART, SUPER-FAMILY, TIGRfam, PIR, PROSITE, and PRINTS (Mulder *et al.*, 2005). Thus, multiple analyses can be performed with a single sequence search that returns domain match scores from the multiple databases (Fig. 3.6). When matches are found, InterPro identifies the domain along with the databases that provide the predicted match. Also provided are the amino acid range that contains the match, the score of the match, and an assignment of whether the domain is considered true, a partial match, unknown, or a false positive. The classifications of true, partial, and unknown matches are assigned by the contributing databases. False positive is a curated designation given by InterPro to contributions.

InterPro shows the e-value for sequences classified as true hits. Two e-values are provided, one for the domain and the other for the sequence. For proteins with a single domain, the e-values are identical. When a sequence has multiple domains, the e-value for the sequence is either the sum (Pfam) or product (PRINTS) of the e-values for the motifs (Karlin and Altschul, 1990). Thus, a protein with multiple domains or motifs in Pfam or PRINTS may have high e-values despite statistically significant individual domain matches.

Information regarding potential or overlapping matches within the query sequence must be found at the individual databases. For example, at PROSITE and Pfam we can obtain information on putative domains. SMART results in InterPro are displayed to users as a set of non-conflicting domains with high scores. Within SMART, low scoring domains and "Hidden" domains that overlap with higher scoring domains can be found. Similarly, contextual domains are shown at Pfam but not in InterPro. Contextual domains whose match score falls below the gathering cutoff but are still considered a match because of the presence of other domains in the sequence are visible at Pfam.

To validate and provide additional evidence for predicted domain or family resemblances, control procedures are performed. (1) We can isolate and re-run the sequence of the match region against the databases. If the matched region contains a domain, a positive and similar result is expected when the sequence has been removed from the other amino acids. (2) We choose a known member of the family and do reverse search for our sequence. If the sequence is a member of the protein family, a family member should be able to retrieve the sequence from the sequence database or from a sequence record within the family-domain database. (3) Use a shuffled version of the sequence as a negative control. Presumably a sequence of the same composition but random order will not retain the domain characteristics. If the same match is obtained, it suggests the

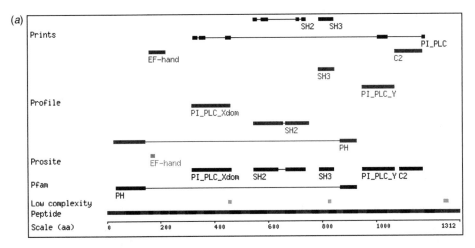

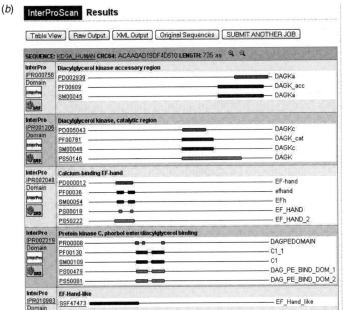

Figure 3.6 *(a) Ensembl and (b) InterPro predicted domain matches from the same query sequence search against the integrated databases. The databases within Ensembl and InterPro are shown to the left with match results on the right. The display of multiple results makes apparent that the domains predicted by one database differ from those predicted by another. It is important to have multiple analyses of your sequence. Two independent methods of identifying the same domain is stronger evidence for the domain being present within the sequence than any single database prediction.*

match is due to a biased composition. Nonetheless, the searches result in computational predictions based on families and domains reduced to defining characteristics, sequence or structure. These features may be found in proteins in which they are not functional as domains. Unless experimental evidence confirms the presence of the domain, the domain is only predicted.

3.4   CONCLUSION

The conservation of protein function occurs both at the level of structure and sequence. The globin family is the classic example of a family of proteins whose structure is highly conserved while the sequences are very divergent. At the other end of the spectrum, serine/threonine kinase motifs are conserved in sequence and function. The pairwise sequence similarity searches employed at sequence databases were insufficient to routinely find with confidence short, highly conserved sequences, consensus patterns or structurally conserved domains dispersed throughout the highly populated sequence databases. Bioinformaticians have improved our ability to detect family and domain members by reducing the overall search space for sequences and utilizing formal definitions for families and domains.

The use of family-domain databases increases specificity and sensitivity for family members and protein domains. This is achieved in part because the dedicated family-domain databases decrease the overall search space for related sequences. Fewer sequences and residues in the database decrease the likelihood of random matches due to the size of the database. The ability to detect distantly related sequences was enhanced through improved multiple sequent alignment algorithms, the development of position specific scoring matrices and use of hidden markov models. These advancements provide both more specific and more sensitive searches for families and domains.

By taking into account the focus of a database (family or domain), how the representations are created (consensus, pattern or profile) and the manner in which the results are returned (all matches, potential matches, etc.), it is possible to create informative searches for known and putative protein functions. The BLOCKS database identifies highly conserved domains. SMART classifies and catalogues multi-domain signaling molecules. PRINTS and PROSITE contain family descriptions that encompass entire sequences. Pfam focuses on families as defined by domains and hidden Markov models. The integrated databases, InterPro and Ensemble, increase the number of possible family and domain matches returned because they query multiple databases in one search. Regardless of whether your matches are returned from searching a single or multiple databases, the evidence is inferred functionality based on family-domain sequence and structures. These suggested functions can then be explored and tested experimentally.

BIBLIOGRAPHY

Attwood TK, Bradley P, Flower DR, *et al.* (2003). PRINTS and its automatic supplement, prePRINTS. *Nucleic Acids Research* 31(1):400–402.

Bairoch A (1992). PROSITE: a dictionary of sites and patterns in proteins. *Nucleic Acids Research* 20:2013–2018.

Baldi P, Chauvin Y, Hunkapiller T, *et al.* (1994). Hidden Markov models of biological primary sequence information. *Proceedings of the National Academy of Sciences USA* 91(3):1059–1063.

Blundell T, Sibanda BL, Pearl L (1983). Three-dimensional structure, specificity and catalytic mechanism of renin. *Nature* 304:273–275.

Bork P, Gibson TJ (1996). Applying motif and profile searches. *Methods in Enzymology* 266:162–184.

Dayhoff MO, Schwartz RM, Orcutt BC (1978). A model of evolutionary change in proteins. In: Dayhoff MO, ed. *Atlas of Protein Sequence and Structure, Vol. 5, Suppl. 3*. Washington, D.C.: National Biomedical Research Foundation; pp. 345–352.

Eddy SR (1998). Profile hidden Markov models. *Bioinformatics* 14:755–763.

Eddy SR, Mitchison G, Durbin R (1995). Maximum discrimination hidden Markov models of sequence consensus. *Journal of Computational Biology* 2:9–23.

Edgar RC (2004). MUSCLE: multiple sequence alignment with high accuracy and high throughput. *Nucleic Acids Research* 32:1792–1797.

Feng D, Doolittle RF (1987). Progressive sequence alignment as a prerequisite to correct phylogenetic trees. *Journal of Molecular Evolution* 60:351–360.

Finn RD, Mistry J, Schuster-Böckler B, *et al.* (2006). Pfam: clans, web tools and services. *Nucleic Acids Research* 34(Database issue):D247–D251.

Grasso C, Lee C (2004). Combining partial order alignment and progressive multiple sequence alignment increases alignment speed and scalability to very large alignment problems. *Bioinformatics* 20(10):1546–1556.

Gribskov M, Veretnik S (1996). Identification of sequence pattern with profile analysis. *Methods in Enzymology* 266:198–212.

Gribskov M, McLachlan AD, Eisenberg D (1987). Profile analysis: detection of distantly related proteins. *Proceedings of the National Academy of Sciences USA* 84:4355–4358.

Henikoff S (1996). Scores for sequence searches and alignments. *Current Opinion in Structural Biology* 6:353–360.

Henikoff S, Henikoff JG (1992). Amino acid substitution matrices from protein blocks. *Proceedings of the National Academy of Science USA* 89:10915–10919.

Henikoff S, Henikoff JG (1994). Protein family classification based on searching a database of blocks. *Genomics* 19:97–107.

Henikoff JG, Henikoff S (1996). Using substitution probabilities to improve position-specific scoring matrices. *Cabios* 12(2):135–143.

Henikoff S, Henikoff JG, Pietrokovski S (1999). Blocks+: a non-redundant database of protein alignment blocks dervied from multiple compilations. *Bioinformatics* 15(6):471–479.

Henikoff JG, Greene EA, Pietrokovski S, *et al.* (2000). Increased coverage of protein families with the blocks database servers. *Nucleic Acids Research* 28:228–230.

Hulo N, Bairoch A, Bulliard V, *et al.* (2006). The PROSITE database. *Nucleic Acids Research* 34:D227–D230.

Karlin S, Altschul SF (1990). Methods for assessing the statistical significance of molecular sequence features by using general scoring schemes. *Proceedings of the National Academy of Sciences USA* 87(6):2264–2268.

Karlin S, Altschul SF (1993). Applications and statistics for multiple high-scoring segments in molecular sequences. *Proceedings of the National Academy of Sciences USA* 90(12):5873–5877.

Krogh A, Brown M, Mian IS, *et al.* (1994). Hidden Markov models in computational biology: applications to protein modeling. *Journal of Molecular Biology* 235:1501–1531.

Lassman T, Sonnhammer ELL (2002). Quality assessment of multiple alignment programs. *FEBS Letters* 529:126–130.

Lesk AM, Chothia C (1980). How different amino acid sequences determine similar protein structures: the structure and evolutionary dynamics of the globins. *Journal of Molecular Biology* 136:225–270.

Letunic I, Copley RR, Schmidt S, *et al.* (2004). SMART 4.0: towards genomic data integration. *Nucleic Acids Research* 32:D142–D144.

Letunic I, Copley RR, Pils B, *et al.* (2006). SMART 5: domains in the context of genomes and networks. *Nucleic Acids Research* 34(Database issue):D257–D260.

Luthy R, Xenerios I, Bucher G (1994). Improving the sensitivity of the sequence profile method. *Protein Science* 3:139–146.

Mulder NJ, Apweiler R, Attwood TK, *et al.* (2005). InterPro, progress and status in 2005. *Nucleic Acids Research* 33(Database issue):D201–D205.

Nagar B, Overduin M, Ikura M, *et al.* (1996). Structural basis of calcium-induced E-cadherin rigidification and dimerization. *Nature* 380:360–364.

Notredame C, Higgins DG, Heringa J (2000). T-Coffee: a novel method for fast and accurate multiple sequence alignment. *Journal of Molecular Biology* 302:205–217.

Saitou N, Nei M (1987). The neighbor-joining method: a new method for reconstructing phylogenetic trees. *Molecular Biology and Evolution* 4:406–425.

Schultz J, Milpetz F, Bork P, *et al.* (1998). SMART, a simple modular architecture research tool: identification of signaling domains. *Proceedings of the National Academy of Sciences USA* 95:5857–5864.

Scordis P, Flower DR, Attwood TK (1999). Fingerprintscan: intelligent searching of the prints motif database. *Bioinformatics* 15(10):799–806.

Sigrist CJA, Cerutti L, Hulo N, *et al.* (2002). PROSITE: a documented database using patterns and profiles as motif descriptors. *Briefings in Bioinformatics* 3:265–274.

Sweet RM (1986). Evolutionary similarity among peptide segments is a basis for prediction of protein folding. *Biopolymers* 25(8):1565–1577.

Tatusov RL, Altschul SF, Koonin EV (1994). Detection of conserved segments in proteins: iterative scanning of sequence databases with alignment blocks. *Proceedings of the National Academy of Sciences USA* 91:12091–12095.

Thompson JD, Higgins DG, Gibson TJ (1994). CLUSTAL W: improving the sensitivity of progressive multiple sequence alignment through sequence weighting, position-specific gap penalties and weight matrix choice. *Nucleic Acids Research* 22(22):4673–4680.

Thompson JD, Gibson TJ, Plewniak F, *et al.* (1997). The ClustalX windows interface: flexible strategies for multiple sequence alignment aided by quality analysis tools. *Nucleic Acids Research* 24:4876–4882.

Thompson JD, Koehl P, Ripp R, *et al.* (2005). BAliBASE 3.0: latest developments of the multiple sequence alignment benchmark. *Proteins* 61:127–136.

Vingron M, Sibbald FR (1993). Weighting in sequence space: a comparison of methods in terms of generalized sequences. *Proceedings of the National Academy of Sciences USA* 90(19): 8777–8781.

Weisstein T, Jungck JR (2006). Evol Seq Biological ESTEEM. BioQUEST Curriculum Consortium. Beloit College. http://www.bioquest.org/esteem/esteem_result.php.

Wright W, Scordis P, Attwood TK (1999). BLAST PRINTS—alternative perspectives on sequence similarity. *Bioinformatics* 15:523–524.

Wu CH, Yeh LS, Huang H, *et al.* (2003). The Protein Information Resource. *Nucleic Acids Research* 31:345–347.

Chapter 4

Getting Started: Modeling

In the next few chapters, we look at models designed to examine the dynamics of biological systems; the changes of the system over time. To some, dynamics may be an obvious domain for models that use continuous time and differential equations. It might be less obvious that discrete models that are stochastic or Boolean in nature may also be used. We focus on the continuous models because the language of change over time and rates is familiar to those of us who have performed temporal studies of our biological topic of interest. In many ways, the next chapters are intended to advance our current understanding of dynamics by introducing new concepts without requiring the acquisition of a whole field of knowledge. With that said, this chapter is an introduction to aspects of numerical modeling with ordinary differential equations. It is meant to be a primer that establishes an intuitive understanding that the following three chapters build on.

The use of mathematical models to explore biological phenomena is conceptually similar to the use of model organisms—*Xenopus*, *Drosophila*, zebra fish, yeast, cell cultures—for experimental research. Model organisms are chosen in order to facilitate discovery of factors and mechanisms that we believe can be applicable beyond the specific organism. We do not expect the results of the studies done in model organisms to exactly reproduce processes in other organisms or cell types. Rather, we look to identify a set of conditions, interacting factors, and organization that reproduce the analogous behavior. The research design often requires simplifications and approximations to be made (Wilmsatt and Schank, 2001). The research design and experimental conditions determine how broadly the results can be interpreted or applied. Experimental conditions are varied in order to determine the underlying structure of the system and its behavior. The same is true of mathematical models.

When modeling biological systems, whether the system is metabolism, transcription, translation, or vesicular transport, the same general cycle of steps are taken (Bower and

A Cell Biologist's Guide to Modeling and Bioinformatics. By Raquell M. Holmes
Copyright © 2007 John Wiley & Sons, Inc.

Bolouri, 2001) (Fig. 4.1). The first step is familiar to biologists and that is the development of the noncomputational or calculation-independent model (Gibson and Mjolson, 2001). This involves creating a concept map that embodies your understanding of the molecular components and processes involved in the biological system. The remaining steps deal with creating the mathematical and computational model, steps commonly performed by computational scientists, modelers, applied mathematicians, or statisticians.

The modeling process that we go through in this book involves making a concept map of your biological system, a set of system statements, symbolic substitutions within model and system statements, creating a list of ordinary differential equations (ODEs), and choosing kinetic models for the reaction mechanisms of the cellular processes. The computational steps are the entering of model components (variables, reactions, rate equations) into a simulation tool and the setting up of simulation parameters (tasks, methods) simulating and visualizing results.

The purpose of one's model guides one's selection of factors, biochemical reactions, and cellular processes included in a conceptual map. These choices then dictate what is possible to explore in the model. The appropriateness of a model will depend on what aspect of the biological phenomenon one is interested in. For example in cell cycle models, if one is interested in the effect of phosphorylation by the cell cycle kinase cdc2 on the kinetics of cell cycle progression, a model that only grossly depicts activation and inactivation of M-phase promoting factor (MPF) would be insufficient. Conversely, if one's focus requires knowing solely that MPF is activated, the detailed dynamics of the binding rates of cdc2 and cyclin may be more cumbersome than helpful (see Chapter 6).

The models in this book are mathematical models described with a system of ordinary differential equations and simulated by providing numerical values. Ordinary differential

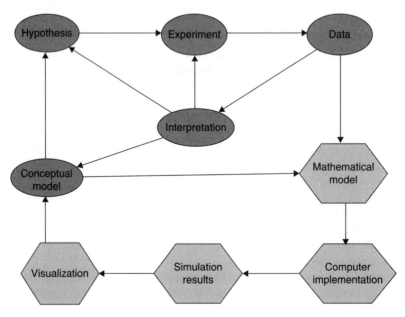

Figure 4.1   *The schematic shows a research cycle with steps common to biologists in circles, and steps performed by computational scientists, modelers, and applied mathematicians/statisticians in pentagons.*

equations are constructed based on our conceptual and theoretical model of the interactions responsible for the cellular behavior of interest. In the next pages, we walk through each step of creating a mathematical model in order to highlight the process. We use the same steps in the subsequent chapters in which the focus is on modeling particular cellular processes and introducing simulation tools that are publicly available.

4.1 CREATING A NONCOMPUTATIONAL MODEL

The noncomputational model answers the questions:

1. What are the biochemical reactions to be modeled?
2. What are the kinetics of the system and reactions?
3. What are the parameters and initial conditions for the system?

These questions are answered with data from one's experiments and the literature with the limiting assumptions one makes based on one's hypothesis and purpose for the model.

4.1.1 What Are the Biochemical Reactions to be Modeled? From Diagrams to Reaction Maps

What are the biochemical reactions? The first steps of creating a model involve identifying exactly what is being modeled. Concept maps in the context of modeling are the schematic or diagrammatic representation of the components and interactions that will be modeled mathematically. We often use diagrams to illustrate the current state of understanding rather than try to use our linear language to describe the nonlinear nature of pathways and process. These diagrams draw components and their interactions to illustrate the structure and directionality of the pathway or biological phenomenon.

Concept maps contain information on the components (i.e., sugars, proteins, enzymes, lipids, metabolites); the flow of mass (i.e., biochemical reactions and membrane transport); and modulators of the flow (i.e., cofactors or catalysts). For example, in the diagram of glycolysis (Fig. 4.2), we can see that the pathway components include 10 carbon-based sugar molecules, 10 enzymes, and 2 types of carrier molecules. The mass—carbon molecules—flows from glucose through a series of intermediates to become pyruvate.

Developing a noncomputational model involves creating and moving between sets of representations, from diagrams to maps and then to mathematical descriptions. The conceptual models developed by cell biologists have common notation—arrows for processes, circles for small molecules, rod structures for receptors in membranes, and so forth. The increased use of commercial products with image-based vocabularies is also contributing to common image-based notation of pathways. However, there is no agreed upon standard for schematic representation of molecular interactions. In contrast, biochemical notation has a standard form for defining biochemical reaction maps. By using biochemical notation standards to create an interaction map, we move one step closer to the mathematical model.

We can redraw reactions from diagrams into a more standard biochemical form (Fig. 4.3). To transition from common biological representations to mathematical ones in the diagram, we assign each molecule acting as substrate, product, or cofactor a

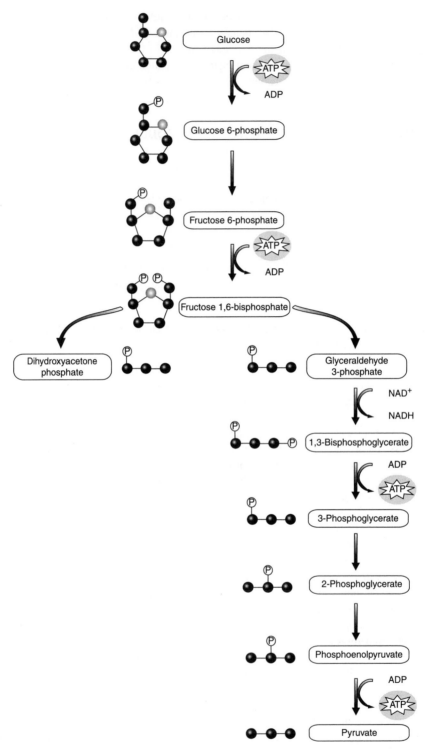

Figure 4.2 *Schematic of the glycolytic pathway using familiar biological terminology and conventional biochemical structure drawing.*

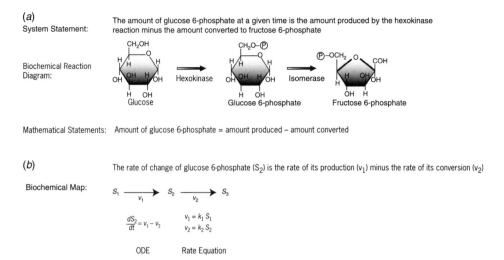

Figure 4.3 *Translating representations. A simple series of biochemical reactions can be described by a system statement, biochemical reaction diagram, or mathematical statement. (a) A system statement and biochemical reaction diagram of the processes including a simple subtraction statement for determining the amount of glucose 6-phospate for a single point in time. The same biochemical reactions, glucose conversion to glucose 6-phosphate and glucose 6-phosphate conversion to fructose 6-phosphate, are shown. (b) The process described in terms of rates of change and with substitution of biochemical notation in the biochemical map for chemical structure drawings. The differential equation and the governing rate equations are shown below.*

symbol (e.g., X or S). The arrow represents a specific reaction and its associated velocity (V). Each drawing is a representation of the same biological process. Both representations convey something different, and each is "correct" or appropriate for its audience. However, for biologists, the representations with names of the metabolites and enzymes may seem more informative because they provide terminology that allows for researchers to draw on knowledge about the molecule that is not explicitly presented in the diagram. In contrast, a symbol's meaning comes solely from the context of the flowchart. Making our biological maps and diagrams use conventional and uniform mathematical symbols brings us a step closer to the conversion of a biological diagram to a mathematical representation.

In a number of modeling articles, the authors refer to a set of system statements that are the basis for their mathematical models. A system statement says what is happening. For example, "Glucose is converted to glucose 6-phosphate" is a system statement. Well-written system statements provide sufficient information to appropriately construct either the concept map or a set of ordinary differential equations. We can use our diagrams and biochemical reaction map as guides for writing system statements and differential equations for the model. Each process in the model is written as a statement. The reaction map and sentences can in turn be translated into mathematical statements by substituting symbols and signs appropriately. In the case of glycolysis (Fig. 4.3), the map indicates that glucose 6-phosphate (S_2) is produced by the phosphorylation of glucose (v_1), and it is removed by its conversion to fructose 6-phosphate (v_2). This is the system statement. It defines how a variable changes in relation to other processes in the model.

4.1.2 What Are the Kinetics? Rate Equations

There are two sets of equations that are developed for our math models: a set of differential equations to describe the system behavior (system equations) and the set of algebraic equations used to describe the rates of change of specific variables (rate laws). The system statement and reaction map can be used to guide writing the differential equation.

For the rate of change of glucose 6-phosphate, the ODE is

$$\frac{dS_2}{dt} = v_1 - v_2$$

This is a shortcut for writing differential equations from concept maps and system statements. It is not the formal description of constructing ordinary differential equations.

If we modify our model of glucose 6-phosphate (S_2) production to include conversion of fructose 6-phosphate back to glucose 6-phosphate (v_3), the differential equation would be written as

$$\frac{dS_2}{dt} = v_1 - (v_2 + v_3)$$

The set of ordinary differential equations are fully described when the velocities are assigned rate equations that define how the process occurs over time. Typically, the rate equations are algebraic equations of familiar enzymatic rate laws. Enzyme kinetics are themselves mathematical models of the behavior of the reaction. The models in the following chapters involve mass action kinetics, exponential decay, and Michaelis-Menten and Hill-type kinetics.

Because the rate equations govern the behavior of the reactions within the model, it is worth taking time to examine the kinetic models that will be used in the following chapter. These are kinetic models that are commonly discussed in biochemistry courses. Readers who have a good understanding of how reaction processes are described with kinetic models may wish to skip to the section on parameter values and implementing the computational models.

4.1.2.1 Kinetic Models
The factors of our biological system are represented as variables and rate equations in our mathematical model. Metabolites, proteins, and ions that are the substrates and products of reactions are variables. These are the molecules to which we assigned the notation of X or S. The variables represent concentrations of molecules that will change over time. Enzymes are included explicitly as variables only when their concentrations change over time. For the models in this book, we assume that the amount of enzyme does not change during the course of the simulation. Given this, the enzymes and reaction mechanisms are implicitly represented as rate equations.

Highlight 4.1

When in doubt, assume enzyme concentrations are constant. This simplifies rate equations and makes it possible to use conventional enzyme kinetic models.

Rate equations are a symbolic representation for the way a variable changes over time (Fig. 4.4). The law of mass action refers to a process being directly proportional to the

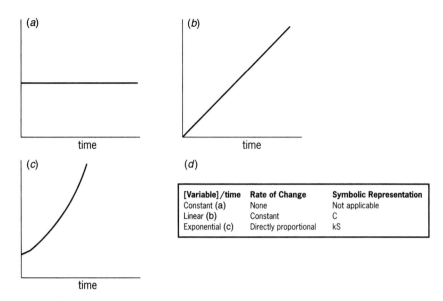

Figure 4.4 *(a) Constant, (b) linear, and (c) exponential behaviors for the amounts of variables with respect to time are shown as graphs. (d) Relationships between the qualitative behaviors of the variables, rate of change, and symbolic representations of the rates are listed across the table rows.*

concentration of the variable. Proportionality in general for two variables is written in the form $y = kS$, and k is the proportionality constant. Similarly, mass action rate equations are written as a rate constant (k) times the concentration of the substrate (S). They hypothesize that the rate of change, commonly denoted as velocity (v), in the variable (S) is proportional to the concentration of substrate and a rate constant (k). Mass action rate equations can also be written for processes involving more than one substrate (Table 4.1).

Chemical notation: $S\text{->}$

$$v = -kS \tag{4.1}$$

$$\frac{dS}{dt} = -kS \tag{4.2}$$

Mass action kinetics are appropriate when the process is mediated by an unregulated enzyme or an enzyme unlikely to be saturated, such as when the velocity of the reactions are linear. To calculate the rate constant from experimental data, we plot the velocity against concentration on a graph. Assuming the data creates a straight line, we determine

TABLE 4.1 Mass Action Rate Equations for Biochemical Reactions

Reaction Mass	Action Equation	Unit of Rate Constant (k)
->	k	$\mu M/s$
$S\text{->}$	kS	s^{-1}
$S_1 + S_2\text{->}$	$kS_1 * S_2$	$s^{-1} * \mu M^{-1}$

the slope, which is equivalent to the rate constant. The change in concentration of the substrate over time for reactions with a linear rate equation is an exponential.

Highlight 4.2

Unregulated enzymes are unlikely to be saturated and can be modeled as a fixed rate constant.

Mass action is appropriate for describing enzyme-catalyzed reactions when the velocity of the reaction is proportional to the substrate concentration. However, not all reactions have a linear relationship between substrate and rates of change as modeled by mass action kinetics. As well, we may want to study reactions near saturating conditions, which requires different kinetic models. More complex models of enzyme kinetics provide us with rate equations that can describe reactions in saturated conditions or involving more than one molecular interaction.

The Michaelis-Menten kinetic model is able to describe the behavior of enzymes when concentrations of substrate are very low or saturating. Michaelis-Menten takes the form:

$$\frac{V_{\max} * S}{K_{\mathrm{m}} + S} \tag{4.3}$$

$V_{\max}$ is the maximum rate of the reaction, which is reached at saturating concentrations of substrate. K_{m} is the substrate concentration at which the reaction is at half its maximal velocity. This model can be reduced to mass action kinetics when the enzyme is unsaturated and the process is irreversible. How is this possible? In cases when substrate concentration (S) is small in relation to K_{m}, the system is unsaturated and we assume that its contribution is insignificant in relation to the much larger value of K_{m}. The formula can thus be rewritten as:

$$\frac{V_{\max} * S}{K_{\mathrm{m}}} \tag{4.4}$$

$$\frac{V_{\max}}{K_{\mathrm{m}}} = k \tag{4.5}$$

$$k * S \tag{4.6}$$

Substituting Eq. (4.5) within Eq. (4.4) gives us the notation for mass action kinetics [Eq. (4.6)].

The two parameters used in the Michaelis-Menten rate equation are the $V_{\max}$ of the enzyme and the Michaelis constant, K_{m}. These values can be determined through *in vitro* studies in which the enzyme concentrations are held constant and the substrate concentration varied (Stryer *et al.*, 2002). By varying the concentration of substrate, we are able to determine the maximum velocity for the reaction ($V_{\max}$) as well as the concentration of substrate at which the reaction occurs at half its maximal speed (K_{m}). These parameters are properties of the enzymes.

Hill type kinetics [Eq. (4.7)] are used when the enzyme, pump, or channel displays cooperative behavior, present in multimeric proteins (e.g., hemoglobin, inositol triphosphate receptor) (Hill, 1910). The original Hill equation described the fraction of sites on a multimeric protein occupied the substrate, given that when the substrate binds to one site it has a positive or negative affect on the binding of subsequent substrate. The Hill coefficient is a term introduced to account for the nonproportional effect of substrate binding on the rate of the reaction. The Hill coefficient (n) is calculated from the experimental data. Positive Hill coefficients reflect an increased enzymatic activity in relation to substrate binding, whereas a negative coefficient represents negative cooperativity.

$$\frac{V_{\max} * [S]^n}{[S]^n + K_{\mathrm{d}}^n} \tag{4.7}$$

4.1.3 What Are the Parameters and Initial Conditions?

Parameters (i.e., K_{m}, k, $V_{\max}$, etc.) are essential values for modeling dynamic processes. Each rate law used to describe how a biochemical reaction or process occurs contains a rate constant or parameter. What are parameters? They are properties associated with the enzyme or biological process and are derived by fitting experimental data to a mathematical model. Parameters include the enzymatic rate constants and system parameters (e.g., conserved mass of adenosine phosphate) that do not change during the simulation but are required to accurately describe the dynamic behavior of a biological process.

The set of parameter values in a given model are often pieced together from different literature reports created by different labs. Even when two labs are characterizing the same enzyme or biological process, variation exists within the parameter values due to differences in conditions under which they were measured, temperature, pressure, and substrate or enzyme concentrations. In many cases, some rate values are known and others are not. In addition to primary literature, parameter values for reactions are being gathered and curated in Web-accessible databases such as Brenda, SigPath, and ProcessDB (Table 4.2).

Highlight 4.3

Identify a small set of parameters for your system based on experiments or literature.

It is not always possible because of unfeasible experiments or lack of data in the literature to obtain a specific rate for one's process of interest. In these cases, an initial estimation is to assume that the K_{m} is approximately equal to the *in vivo* concentration of the substrate. Why? The cell is rarely in a saturated state or far from equilibrium. This suggests that the process occurs within a substrate concentration range in which the rate

TABLE 4.2 Database Sources for Parameter Values

Database	Web Site
BRENDA	http://www.brenda.uni-koeln.de
SigPath	http://icb.med.cornell.edu/crt/SigPath/index.xml
ProcessDB	http://www.integrativebioinformatics.com/processdb.html

of production is responsive to the addition or removal of substrate. This occurs when reactions are at their K_m, they are at half their maximum velocity. This is the linear portion of a reaction that is plotted as velocity (y-axis) versus substrate (x-axis).

Highlight 4.4

K_m of an enzyme that shows saturation kinetics is assumed to be close to the concentration of the substrate *in vivo*.

Even with highly characterized systems, parameter values may be missing. This creates an opportunity for mathematical modelers to estimate values (Zwolak *et al.*, 2005) and for experimentalists to determine them experimentally. Microscopic techniques have been used to determine rates of diffusion, complex formation, and vesicle trafficking (Lippincott-Schwartz *et al.*, 2001). Additionally, radioactively tagged ions have been used to monitor rates of molecule transport by channels or pumps (Lippincott-Schwartz *et al.*, 2001). The aforementioned methods commonly have the advantage of being used with live cells and can obtain fairly high resolutions that lend themselves to kinetic studies. Parameters can also be obtained from samples fixed over a time course, which allows us to graph the change of a variable over time. In the subsequent chapters, we make use of the kinetic models and parameters developed by biochemists and used in existing computational models.

4.1.3.1 *Initial Conditions of Variables*   Initial conditions—parameter values and variable concentrations—are the final pieces of information about our biological systems needed to implement our mathematical model as a computational model. We addressed obtaining parameter values above.

In mathematical models, we consider two main types of variables: dependent and independent. In our models, both molecular concentrations and time change, both are variables. However, time proceeds whether or not there is a change in substance. Time is an independent variable. In contrast, if there is no change in time, there is no change in the concentration of the molecule. Change in the concentration of molecules is dependent on time. These are the factors in our model that we denoted as an X or S.

The variables that change in relation to time during the simulation are referred to in modeling literature as system variables, state variables, or dependent variables. Each is useful for highlighting the variable's relationship to the model. A system variable varies within the system of equations. State variable: calcium exists in two states, bound and unbound. Dependent variable: the change of S is dependent on change in time. Time and space are typically, but not exclusively, independent variables.

In the following chapters, calcium, glucose, and MPF are dependent variables in the models being discussed. Calcium and MPF exist in more than one state. The multiple calcium states are due to its distinct cellular locations. Cytosolic calcium is a different state than calcium in the endoplasmic reticulum (ER), and bound calcium is distinct from free calcium. MPF is both active and inactive. Each variable state participates in a separate set of reactions. We will monitor how each state variable, which correspond to molecular states, changes through the course of a simulation.

Although concentrations for molecules may vary in cells based on the cellular state (quiescent, proliferating, differentiated, progenitor, etc.), one concentration value must be assigned to the variables to run a simulation. Based on our understanding of the biology and biochemistry of the system, we can choose a range of values for a given variable that encompasses the average concentration found *in vivo*. We subsequently examine how the different concentrations affect the simulation. A simulation that looks reasonable for only one concentration value is severely handicapped and unlikely to be representative of a realistic scenario within a dynamic cell.

Detail 4.1

Determine sensitivity of the model to parameters by incrementing up and down from the initial value. Use automated or semiautomated scanning parameter functions when available.

Initial concentrations are simultaneously significant and immaterial. The behavior of the models we create is governed by the ordinary or partial differential equations. Initial concentrations tell the approximation process where to begin to determine a solution. By definition in systems of ODEs and partial differential equations (PDEs), for each set of initial conditions there is a unique and consistent solution. This makes each concentration significant. Yet, if we were to solve the equations symbolically, using only the relationships defined in the rate equations and the system of ODEs, no specific number value is given. Think of our enzyme kinetic models, which are symbolic representations of the enzyme kinetics. We know that the plot of the velocity versus substrate concentration of a Michaelis-Menten model will create a hyperbolic curve, even without plugging in specific numbers. It is the parameter values and initial concentrations that add numerical values to the graph, but it is the symbolic representation, the algebraic equation that gives the curve its form.

4.2 COMPUTATIONAL MODEL: SIMULATION METHODS

In the chapters that follow, the software tools employed write and solve the differential equations of the modeled system for us. It is important, however, to have an understanding of what issues arise when differential equations are solved on a computer and to understand the effect that the approximation method used to solve the equations has on model results. These topics generally fall under computational science or simulation methods. There are a number of texts dedicated to these topics or containing accessible introductory chapters (Heath, 2001; Shiflet and Shiflet, 2006). We will discuss solving differential equations briefly here to provide reference materials for the future chapters and investigations.

4.2.1 ODE Essentials

We use ODEs to describe the change of variables in relation to time; rates of change. In general, ODEs examine the change in a variable in relation to a single independent variable. In our models, we use them to look at the change in dependent variables, MPF, calcium, cyclin, and so forth, in relation to the independent variable of time. Differential equations use the current state of the model—numerical values of the parameters and concentrations—to determine the next state of the system. The reaction equations and

ODEs govern the behavior of the system. Changing any given parameter or initial concentration value creates a different system state that has its own unique solution. Numerical integration methods (see text below) are used to obtain the solution to the ODEs that govern the behavior of the model.

4.2.2 Approximation Methods and Time Steps

To solve the set of differential equations, numerical approaches are employed that approximate the solutions. In other words, we fully describe one set of initial conditions, concentrations at time zero, and approximate the solution of the differential equations that governs the resultant behaviors. This is what is meant by numerical simulation. It is a simulation in which specific values are provided for each variable and parameter in order to solve the set of differential equations that describe the biological system.

We use simulation tools that have built-in approximation methods. Multiple numerical integration methods have been developed (Table 4.3). A seemingly easy to understand method of approximation is the Euler method. The Euler method is a finite difference equation that resembles a simple subtraction or addition equation. It can be described as stating, "What you have is equal to what you had plus what changed."

The Euler method determines the specific value of what you have at any given time through a series of time steps, delta t (Δt), of predetermined and fixed length (Fig. 4.5). This is similar in some ways to a time series experiment in which samples are taken at a set time interval (i.e., every 5 minutes) to determine the concentration of a variable (e.g., calcium, cyclin, etc.). The Euler method takes discrete steps of Δt in time and plugs the existing variable concentrations back into the governing equations to calculate the next set of concentrations. The size of time steps is critical to the final solution obtained.

4.2.2.1 Approximation Errors: Blowing Up and Rounding Off It is not needed for us to become full experts in the methods of approximation. However, we do need to have a qualitative understanding of the pitfalls and strengths associated with approximation methods. As well, we need to be able to identify the characteristics of a flawed simulation due to the approximation method. There is an inherent limitation to numerical approximations that must be taken into account, which is the error between the approximate solution and the exact solution of the differential equations. Again, Euler is a great example. For each time step (Δt), there is a possible error that is equivalent to Δt^2. For fourth order Runge-Kutta, another method of approximating differential equations, the equation for approximating the solution of the ODE is more complex, but the error range is Δt^5. Thus for a very small Δt, fourth order Runge-Kutta is less vulnerable to approximation errors than Euler. Other approximation methods employ a

TABLE 4.3 Numerical Integration Methods

Software	Fixed Time Step	Stella	Gepasi	Virtual Cell
Euler forward	X	X	X	X
Runge-Kutta 2	X	X	X	X
Runge-Kutta 4	X	X	X	X
Newton	—	—	X	—
Adams	X	—	X	X
BDF	—	—	X	—
LSODA	—	—	X	X

Euler

(a)

population (t) = population ($t - dt$) + growth ∗ dt

growth = rate constant ∗ population

dt = time step

The population at one time step, population (t), is the population from the previous time step, population ($t - dt$), plus the estimated change in population for that time step, growth ∗ dt. Growth is proportional to the population.

(b)   (c)   **Cell Population Growth**

	time: days	population (t)=	population (t-dt)	growth	delta t
$\xrightarrow{\hspace{1cm}} S$	0.00	10			
v_1	0.25	12.5	10	10	0.25
	0.50	15.625	12.5	12.5	0.25
$v_1 = k_1 * A$	0.75	19.53125	15.625	15.625	0.25
rate constant = k_1 = 1/day	1.00	24.41406	19.53125	19.53125	0.25
initial A = 10	1.25	30.51758	24.41406	24.41406	0.25
	1.50	38.14697	30.51758	30.51758	0.25
$S(t) = S(t - dt) + v_1 * dt$	1.75	47.68372	38.14697	38.14697	0.25
$v_1 = k_1 * S$	2.00	59.60464	47.68372	47.68372	0.25
	2.25	74.50581	59.60464	59.60464	0.25
dt = time step	2.50	93.13226	74.50581	74.50581	0.25
	2.75	116.4153	93.13226	93.13226	0.25
	...	...	...	...	...

Figure 4.5 Euler equation. The Euler equation is known as a finite difference equation. (a) We show the equation of change in a cell population at one time (t), using mathematical notation and English. (b) Population growth (S), whether cells, an enzyme or organism, is drawn in biochemical notation. (c) The results of calculating the change in population based on the Euler equation with a Δt of 0.25 are shown as a table. Replace the word population with S to obtain biochemical notation in table.

variable time step such that the most appropriate time step to avoid approximation errors is selected by the algorithm throughout the approximation.

This is particularly true for Euler and Runge-Kutta methods that use fixed time steps set by the user to approximate solutions to the set of ODEs. If the time step is large relative to the time it takes for biochemical reactions to occur within the simulation, the behavior is erroneously described. In the hypothetical system of cell population growth (Fig. 4.5), we have a cell population that has a flat rate of growth: 10 cells per day are produced. Cell removal from the population is proportional to the number of cells in the population. For this model, we have varied the rate constant for the removal of cells. The time step chosen for the approximation with Euler method is 1 day. However, something unintuitive occurs. When the rate constant is set to 5/day, we see sharp oscillations in the results (Fig. 4.6). If we looked at the numbers, we would see that the system is varying sharply between 20 and 0. The same system can be approximated with a time step of 0.25 or 0.125 days. The time step for the approximation is set by the modeler when using fixed step methods such as Euler and Runge-Kutta. The smaller time step results in a smooth curve for the model system.

Again, we can draw a parallel between experimental and simulation time series. The biological behavior observed experimentally in a time series experiment can be dependent on the interval of sampling. Smaller time intervals taken over the course of the biological behavior provide a better overall understanding of the system. A general rule for selecting time step size is to choose a Δt that is one over the smallest timescale. If the smallest time constant is 10 seconds, the Δt would be 0.1 seconds.

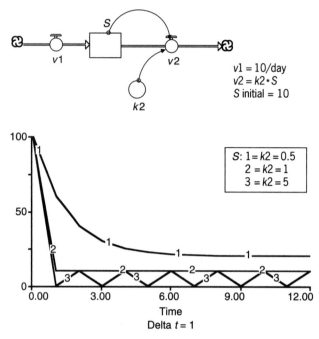

$v1 = 10/\text{day}$
$v2 = k2 * S$
S initial $= 10$

S: $1 = k2 = 0.5$
$2 = k2 = 1$
$3 = k2 = 5$

Time
Delta $t = 1$

Figure 4.6 *Erroneous behavior due to Δt approximation error. The conceptual diagram (drawn in Stella) indicates a population that is fed by an input (v1) and removed by an output (v2). The input occurs at a flat rate of 10/day. The output is proportional to the population (S). The plotted results show the calculated growth of the population with different rate constants (1:0.5, 2:1, 3:5) and the same Δt, 1. The behavior of the system is smooth except for the simulation with a rate constant of 5 and Δt of 1. Note the sharp oscillatory behavior. This behavior is obtained when Δt is relatively large in relation to the overall rate of the simulated process. The same rate constant values simulated with a Δt of 0.25 do not show this behavior.*

A characteristic pattern of a model with an inappropriate step size or approximation method is when the graphed simulation results produce oscillations whose peak values increase rapidly over time (Fig. 4.6). This is referred to as "blowing up." When variable values blow up, it is fairly certain that your method of approximation is not working.

Highlight 4.5

Start with a Δt to be 10-fold smaller than the smallest rate constant in the simulation.

There are two approaches to determining whether one's simulation has such sharp oscillations or is blowing up. First, if one is using a fixed step method (Euler; Runge-Kutta), vary the time step by a factor of 2 or 3. Second, choose another method to simulate the model and again run two different time steps. When the same behavior is produced with two time steps that differ by twofold and two different approximation methods, one's model and approximation methods are fine.

Detail 4.2

Try different step sizes and different approximation methods.

4.2.3 Simplifying the Set of ODEs

A common step taken in creating numerical models is the reduction of the number of ODEs or parameters that need to be solved or provided, respectively. There are a few common practices for achieving this: establishing algebraic relationships whenever possible by defining velocities or variables in terms of one another. An overly simplistic example of decreasing the number of differential equations is the identification of variables that are conserved in the system.

In Chapter 5, where we look at a model of glycolysis, the total amount of adenosine nucleotides is conserved. Adenosine nucleotides (A_T) are distributed between ATP (A_3) and ADP (A_2), but the overall amount remains the same in the model.

$$A_T = A_3 + A_2$$

Similarly in Chapter 6, the total amount of MPF is distributed between the two states of active and inactive MPF:

$$MPF_T = aMPF + iMPF$$

A set of variables may be converted to parameters when they exist as conserved mass in the model. The concentrations of each variable (ATP, ADP) change in relation to each other during the simulation but the total concentration (A_T) is constant throughout the simulation. The recognition of conserved variables enables us to solve for one variable algebraically in terms of the other instead of using a differential equation.

$$A_3 = A_T - A_2$$

This removes the differential equation from the set of differential equations to be solved and reduces the number of possible errors due to approximation.

Another common approach to reducing the mathematical complexity of a model is to scale out a term such that the number of parameters required to solve the equation is reduced. Scaling is one of those methods we learned early in our education where we use a fraction equivalent to the value of 1 as a multiple to change the scale of a system although the overall equation is not changed. Simple conversion examples are across scales of concentration, size, or time: nano to micro (moles, meters, or seconds). It is also possible to scale out units. When models are run in the absence of units, it is referred to as dimensionless. These work on the same principle of scaling but find conversion relationships within the model itself.

4.3 VISUALIZING AND ANALYZING SIMULATION RESULTS

An important aspect of analyzing your simulation data is having a well-organized format for storing and organizing the results. The smallest computational model that we examine contains five components and nine parameters. Although in the text we describe one set of parameter values to examine as a case study, each parameter value can be changed to test how the model behaves. If each parameter were tested in two conditions, there would be a minimum of 2^9 (512) test results.

The breadth of possible variations and quantity of data output are part of the amazing strength of developing models. Simultaneously, it creates a wealth of data that can be as daunting or uninformative as the biological system itself. For this reason, it is important to both have an understanding of what one "expects" of the system, what relationships one wants to examine, and the limitations of one's model given the assumptions made during its construction.

4.3.1 Steady-State Analysis

Another test for the stability of the model is to *set* the initial conditions for steady state and run the simulation. Under these conditions, variable concentrations are not expected to change as seen in Figure 4.7*a*. Although the variable concentrations do not change, biochemical reactions are still occurring. S continues to be produced (v_1) and removed (v_2) from the system such that the total amount of S converted by the reactions is not constant but rather increases over time. The rates of S production and removal are equivalent such that the concentration of S over time appears constant. The biological system is in a dynamic equilibrium. When we run a steady-state simulation, we are looking for the set of values that result in the change in variables over time being equivalent to zero.

If a steady state is not reached when simulating a set of ODEs for a biological system that is known to have a steady state, the equations do not accurately model the system. If your system does not reach equilibrium, it may be that the parameter values are off; other factors need to be taken into account either in the rate equations or as additional variables within the model. Determining steady-state conditions makes it possible to change single parameters and examine the impact of these parameters relative to the steady-state behavior of the simulation. If the model is highly sensitive to parameter values, it is referred to as a poorly conditioned system. A slight change in one numerical value causes the system's overall behavior to change drastically. The models we look at in this text are well-conditioned systems.

What is the goal of the model? We will see in the following chapters that the model of the cell cycle reproduces oscillation behaviors but is not exact in the timescale when compared with experimental values. The model was used to examine the plausibility of generating oscillations, not the validation of particular experimental values. In contrast, in Chapter 7, the experiments performed were used to determine factors contributing to the production of experimental results. The temporal and quantitative changes in concentrations of variables are closely reproduced. The expected results of the two models are quite different. In one case, we are looking in the graphs or tables for the presence of an overall pattern; in the other case, quantitative values. It is up to the author of each model to have an understanding of the types of results he or she is expecting.

Many software tools, differential equation solvers, have been developed for the analysis of biochemical reactions and pathways that provide sets of predefined functions that are commonly used by biochemists (Alves *et al.*, 2006). We use these tools as simulators to examine the behavior of the glycolytic pathway and calcium dynamics. We use a general simulator, Stella, to examine the cell cycle. Other simulators such as MATLAB and Mathematica have greater capacity to handle complex models and solve analytical, nonnumerical, models. These tools are nonspecialized in that they are used in many disciplines. We have chosen to introduce a new tool in each chapter. The goal of the chapters is to introduce both the construction of the model and the simulation tool.

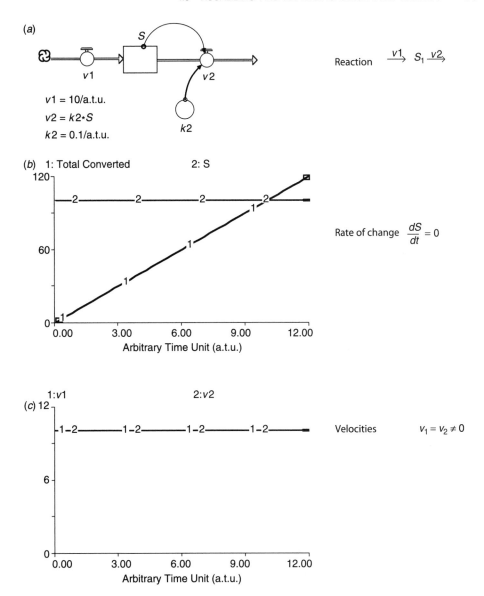

Figure 4.7 Steady state. (a) Simple model consisting of two reactions, v1 and v2, created in Stella. (b) Graph of the total mass converted by the reactions over time (1) and changes in the amount of S over time (2). Although the amount of S is constant, the amount of product created over time increases. (c) The rates for each process (v1 and v2) are equivalent.

Each tool has a different user interface design and functionalities. The tools have easy to understand user interfaces. All of the tools provide simple graphing tools that allow us to view the results as a 2D plot. The software tools with which we will be working vary in the amount of data storage provided, the form of data presentation, and tools for analysis. The simplest form of data is the generation of files either as comma delimited files (Gepasi) or table format (Stella). Virtual Cell supports visualization of temporal data in 2D and 3D spatial models (geometries). As one develops his or her own models and obtains a better understanding of his or her computational needs, one may wish to explore additional tools.

BIBLIOGRAPHY

Alves R, Antunes F, Salvador A (2006). Tools for kinetic modeling of biochemical networks. *Nature Biotechnology* 24(6):667–672.

Bhalla US (2001). Modeling networks of signaling pathways. In: De Schutter E, ed. *Computational Neuroscience: Realistic Modeling for Experimentalists*. Boca Raton: CRC Press; pp. 25–48.

Bower JM, Bolouri H, eds. (2001). *Computational Modeling of Genetic and Biochemical Networks*. Cambridge, MA: MIT Press; 336p.

BRENDA: The Comprehensive Enzyme Information System (2006). Available at www.brenda.uni-koeln.de.

Gibson M, Mjolsness E (2001). Modeling the activity of single genes. In: Bower JM, Bolouri H, eds. *Computational Methods in Molecular Biology*. Cambridge: MIT Press; pp. 3–48.

Heath M (2001). *Scientific Computing: An Introductory Survey*, 2nd edition lecture notes. Available at http: //www.cse.uiuc.edu/heath/scicomp/notes.

Hill AV (1910). The possible effects of the aggregation of the molecules of haemoglobin on its dissociation curves. *Journal of Physiology* 40:iv–vii.

Institute for Computational Biomedicine (2005). SigPath project. Available at http://icb.med. cornell.edu/crt/SigPath/index.xml.

Integrative Bioinformatics (2006). ProcessDB. Available at http://www.integrativebioinformatics. com/processdb.html.

Lippincott-Schwartz J, Snapp E, Kenworthy A (2001). Studying protein dynamics in living cells. *Nature Reviews* 2:444–456.

Michaelis L, Menten M (1913). Die Kinetik der Invertinwirkung. *Biochemische Zeitschrift* 49: 333–369. The current derivation has been proposed by Briggs and Haldane. Briggs GE, Haldane JBS (1925). A note on the kinetics of enzyme action. *The Biochemical Journal* 19:339–339.

Poshusta R (1996). Chemical kinetics. Available at www.sci.wsu.edu/idea/ChemKinetics/.

Shiflet A, Shiflet G (2006). Computational science: scientific programming: differential calculus. Available at http://wofford-ecs.org/ScientificProgramming/DifferentialCalculus/index.htm.

Stryer L, Berg J, Tymoczko J (2002). *Biochemistry*, 5th edition. New York: W. H. Freeman; 1100p.

Voit E, Ferreira A (2000). *Computational Analysis of Biochemical Systems: A Practical Guide for Biochemists and Molecular Biologists*. Cambridge: Cambridge University Press; 544p.

Wimsatt W, Schank J (1998). Modeling—A Primer. In: Jungck JR, Vaughn V, eds. *The BioQUEST Library, Volume 5*. San Diego: Academic Press; pp. 1–12.

Zwolak JW, Tyson JJ, Watson LT (2005). Globally optimized parameters for a model of mitotic control in frog egg extracts. *Systems Biology, IEE Proceedings* 152(2):81–92.

Chapter 5

Modeling Metabolism

The term *metabolic modeling* has been used to refer to two very different research goals that involve metabolic pathways. One goal, common to databases such as BioCyc, KEGG, and Ligand, is the ability to determine the factors and topology (layout or structure) of a metabolic pathway. These databases elucidate which possible genes and proteins are involved in any given set of reactions. The sites use computational approaches to generate a probable or existing metabolic pathway from genomic, RNA expression, and protein interaction data.

Metabolic modeling is also used to refer to creating quantitative models of specific reactions within metabolic networks. These models may be used to predict the behavior of already identified reactions, discover discrepancies between the behaviors of a proposed mechanism and the experimental biological system, to test perturbations of mechanistic explanations, and to better understand the biology of the overall network. In this chapter, we examine the latter case—modeling kinetic reactions in metabolic pathways.

We will focus first on defining the topology of the metabolic pathway (glycolysis), defining the involved components, and writing symbolic representations (see Chapter 4). This will create a noncomputational model. We then focus on developing mathematical descriptions and using a software package, Gepasi, to examine the kinetics of the pathways. Gepasi has been developed for the analysis of biochemical reactions and pathways and therefore provides sets of predefined functions that are commonly used by biochemists. We are using the tool as a general simulator to examine the behavior of the glycolytic pathway and therefore will not be discussing the functions of metabolic control analysis, optimization, or fitting that can be performed with Gepasi.

A Cell Biologist's Guide to Modeling and Bioinformatics. By Raquell M. Holmes
Copyright © 2007 John Wiley & Sons, Inc.

5.1 CREATING A NONCOMPUTATIONAL GLYCOLYSIS MODEL

For this chapter on metabolism, we are using glycolysis to become familiar with creating a model and implementing kinetic simulation models. Glycolysis is one of the first biochemical and metabolic pathways learned by biology students. This is a series of biochemical interactions that break down glucose to produce pyruvate, ethanol, ATP, and NADH. A diagram of this pathway is shown in Figure 5.1. The diagram provides us with a map that we can interpret better than we can describe.

Developing our noncomputational model involves creating and moving between one set of representations to another, from diagrams to maps and then to mathematical descriptions. This is done by addressing the following questions in the context of a research hypothesis:

1. What are the chemical reactions to be modeled?
2. What are the kinetics of the system and reactions?
3. What are the parameters and initial conditions for the system?

5.1.1 Yeast Glycolysis and Oscillations

Scientists have made computational models of glycolysis to examine the kinetics and mechanisms of glycolysis and glycolytic oscillations in yeast (Richter 1974; Richter et al., 1975; Bier et al., 1996; Hynne et al., 2001). Glycolytic oscillation refers to the concentrations of metabolites in the pathway rising and falling with an observable frequency, amplitude, and phase. Glycolytic oscillations in yeast are dependent on rate of glucose addition, pH, temperature, and protein concentration (Richard, 2003; Poulsen, 2004). Oscillations are most obvious at a macroscopic level, i.e. when they are synchronized in the yeast cell population (Das and Busse, 1985; Richard, 2003). Oscillations occur within individual cells (microscopic) and are required for macroscopic oscillations (Aon et al., 1992; Poulsen, 2004). Oscillations can be found in yeast under aerobic and anaerobic conditions in extracts and intact yeast (Chance et al., 1964a, b; Ghosh and Chance, 1964; Hess and Boiteux, 1968; Richard et al., 1993; Richard, 2003; Poulsen et al., 2004).

Why is this particular aspect of biology of interest? One reason is that this biochemical network stays predominately at steady state. It is difficult, therefore, to differentiate degrees of control that each intermediate step has on the system (Olivier and Snoep, 2004). When the system is perturbed into an oscillating state, it is easier to examine the influence of each step on the phase, frequency, or amplitude of the system's behavior. Yeast glycolytic oscillations have been a biological model system for understanding control of steady-state metabolic processes, developing new theoretical methods for analyzing oscillatory systems, and studying intercellular communication (Goldbeter and Lefever, 1972; Bier et al., 1996; Bier et al., 2000; Reijenga, 2001). In this chapter, we look at glycolytic oscillations to obtain a better understanding of the mechanisms that maintain microscopic oscillations within the glycolytic pathway. The models we re-create based on Wolf et al. (2000) investigate the hypothesis that oscillations in ATP are sufficient to maintain oscillations within the glycolytic pathway and NADH when oscillations in particular sugar metabolites in the pathway backbone are absent. As we will refer to the paper by Wolf et al. (2000) frequently throughout our discussion of metabolic modeling, unless otherwise indicated, "Wolf" references this particular paper.

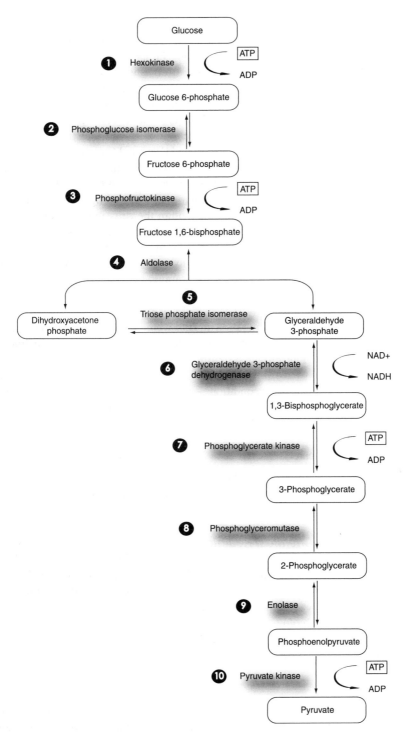

Figure 5.1 *Diagram of glycolysis reaction pathway. The diagram provides both graphical and textual representations for sugars and coenzymes. Text is used to provide names and graphics to identify sugar metabolites. Arrows indicate the flow and direction of mass through the reactions.*

Glycolytic oscillations can be observed in intact cells indirectly by measuring fluorescent absorption by NADH or directly by measuring concentrations of metabolites including NADH over time (de Koning and van Dam, 1992; Richard et al., 1993). Richard et al. (1996a), used the latter procedure to investigate mechanisms for macroscopic oscillations. In these experiments, they detected oscillations in the concentrations of the first three metabolites, glucose 6-phosphate, fructose 6-phosphate, and fructose 1,6-bisphosphate, but none of the subsequent sugar metabolites. This finding differed from previous reports on transient oscillations in yeast extracts in which the concentrations of the downstream three-carbon sugars also oscillated (Betz and Chance, 1965). It was also surprising given NAD reduction and NADH oxidation occur in the reactions producing nonoscillating sugars. Richard et al. (1994, 1996) proposed that ATP hydrolysis and oscillations linked through the glyceraldehydes 3-phosphate dehydrogenase reaction were sufficient for driving and sustaining oscillations within the cells in the absence of oscillations in three-carbon metabolites (Richard et al., 1996). "Richard" shall refer to the paper by Richard et al. (1996) throughout the rest of this chapter.

Wolf created a minimum model of glycolysis to explore whether NADH oscillations could occur in the absence of downstream oscillations and therefore, as proposed by Richard, it was plausible that energy carriers were the mechanism of propagating oscillations through the metabolic pathway. The model was not used as a predictive or quantitative tool. In other words, the simulation results were not expected to quantitatively match the concentration values observed experimentally. Rather, the model qualitatively reproduces behaviors of the anaerobic system.

The model is based on what is known about the biology—rate constants and enzyme kinetics of reactions under anaerobic conditions—and acceptable assumptions (equilibrium states). Wolf evaluated the accuracy of the model in relation to the relative amplitudes and phases found by Richard's experiments in which,

1. ATP/ADP oscillations are greater than NADH/NAD oscillations.
2. Relative amplitudes of upstream sugars are smaller than downstream oscillations.
3. Relative amplitude of ATP/ADP are greater than downstream oscillations.

5.1.1.1 What Are the Chemical Reactions?
To see the choices made by Wolf in creating this model, it is helpful to compare the created model structure to what is known about the pathway. To do this we will use the Kyoto Encyclopedia of Genes and Genomes, or KEGG (Kanehisa, 1997; Ogata et al., 1999; Kanehisa and Goto, 2000; Kanehisa et al., 2006; available at http://www.genome.ad.jp/kegg/) database as a baseline set of known reactions. Although many biologists are familiar with the glycolysis pathway, one might use a metabolic pathway database such as KEGG as a reference (Fig. 5.2) to make a list of pathway reactions and associated enzymes specific to yeast (Table 5.1).

Detail 5.1

KEGG is a database of known and predicted pathways. BioCyc is another metabolic pathway database that could be used. Both databases include information on newly identified genes and therefore are not always accurate. However, the glycolysis pathway in yeast is well documented, and little error is expected. The pathway can be compared with that in BioCyc as a method of confirmation.

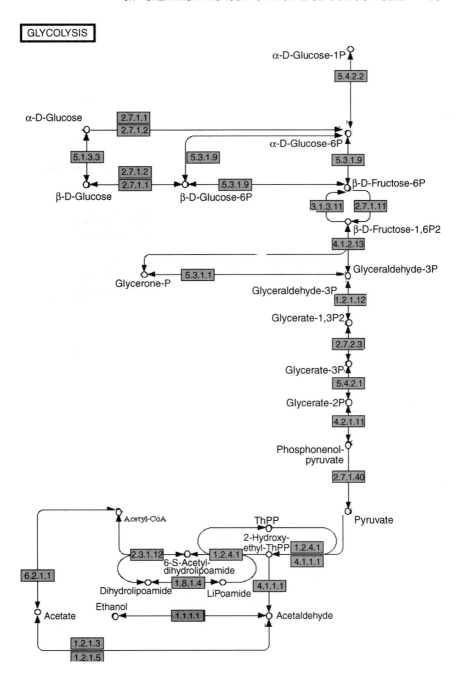

00010 3/23/06

Figure 5.2 *Modified KEGG pathway map of glycolysis. The Kyoto Encyclopedia of Genes and Genomes (KEGG) contains graphical maps of the pathway of glycolysis. The pathway provided in KEGG contains greater detail than is presented here. The image shown allows us to identify the sugars and enzymes involved in the glycolytic pathway for Saccharomyces cerevisiae.*

TABLE 5.1   Information Derived from KEGG Pathway Diagram

Reactions	Enzyme	EC Number
In flux	Glucose	None
Glucose -> Glucose 6-phosphate	Hexokinase	2.7.1.69
Glucose 6-phosphate <-> Fructose 6-bisphosphate	Phosphoglucose isomerase	5.3.1.9
Fructose 6-phosphate -> Fructose 1,6-bisphosphate	Phosphofructokinase	2.7.1.11
Fructose 6-phosphate <- Fructose 1,6-bisphosphate	Fructose-bisphosphatase	3.1.3.11
Fructose 1,6-bisphosphate <-> Glyceraldehyde 3-phosphate	Aldolase (triose phosphate isomerase)	4.1.2.13
Glyceraldehyde 3-phosphate <-> Glycerate 1,3-phosphate	Glyceraldehydes 3-phosphate dehydrogenase	1.2.1.12
Glyceraldehyde 3-phosphate -> Glycerol pathway	NAD-dependent aldehyde dehydrogenase	1.2.1.3
Glycerate 1,3-phosphate -> Glycerate 3-phosphate	Phosphoglycerate kinase	2.7.2.3
Glycerate 3-phosphate -> Glycerate 2-phosphate	Phosphoglycerate mutase	5.4.2.1
Glycerate 2-phosphate -> Phosphoenol pyruvate	Enolase	4.2.1.11
Phosphoenol pyruvate -> Pyruvate	Pyruvate kinase	2.7.1.40
Pyruvate -> Acetaldehyde	Pyruvate decarboxylase	4.1.7.1
Acetaldehyde -> Ethanol	Alcohol dehydrogenase	1.1.1.1

If we compare the set of reactions from KEGG and the Wolf model (Fig. 5.3), we can see that the number of reactions drawn for Wolf are fewer than those drawn for KEGG. Wolf created "lumped" reactions that coupled reactions mediated by hexokinase and phosphofructokinase into one (v_1, Fig. 5.3). Additional lumped reactions combined phosphoglycerate mutase enolase, and pyruvate kinase reactions into one (v4), and all intermediate glycerol-producing steps were reduced to one reaction. Lumped reactions effectively treat a number of reactions as a single reaction by defining them under conditions in which they are equivalent or by describing one reaction in terms of another. In the case of hexokinase and phosphofructokinase, the end products of both hexo- and phosphofructokinase are considered to be at equilibrium and the enzymes insensitive to product concentration (Heinrich and Rapoport, 1975; Teusink and Westerhoff, 2000). The conditions used to create the lumped reaction are described in the methods of the original paper. The simplifications effectively reduced the number of variables used in the calculations.

Related to the question of "What are the chemical reactions?" is the question "What are the components: metabolites, enzymes or cofactors?" We take the components directly from the list of reactions to generate a reference for symbols that we will use through the remainder of the tables and figures (Table 5.2). If reactions were written without explicitly identifying cofactors that we wish to account for in the model, we add them to this list. Because we are examining the role of ATP/ADP and NADH/NAD levels in glycolytic oscillations, they are important components to include. We will not monitor the concentrations of glycerol and ethanol. They serve as molecular sinks for mass moving through the pathway. Their inclusion in the concept map helps us to think about the entire pathway and reactions, but no reactions use these molecules as a substrate.

We arrange the components and interactions into a "flow" map (Fig. 5.4). This map looks very much like Figure 5.3. Drawing the map in "proper" biochemical notation (Voit and Ferreira, 2000) will facilitate our developing the mathematical equations that describe the relationships in the pathway map.

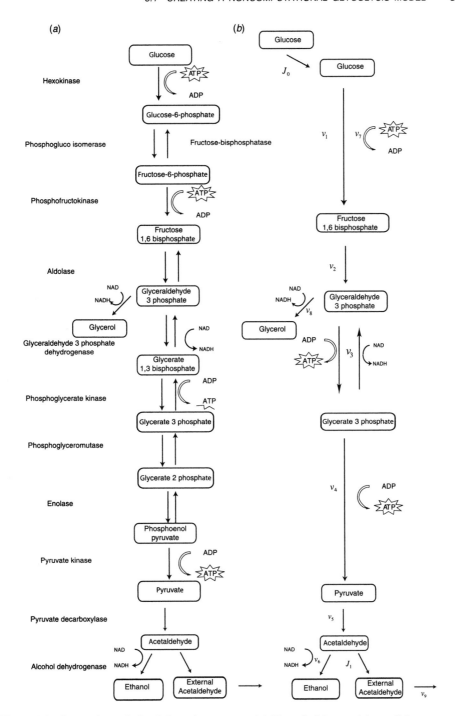

Figure 5.3 *Comparison of glycolytic pathway maps. (a) Map of all known intermediate sugar and enzymes between glucose and ethanol production. (b) Reduced set of intermediates modeled by Wolf et al. (2000) due to lumped reactions. Reactions were lumped together by using known equilibrium ratios to scale out a variable. The unseen intermediates are included in the implicit definition of enzyme rate kinetics and system parameters. Biological names have been used in the image to facilitate the reader's familiarity with the pathway before turning to symbols. The enzymes catalyzing the reactions are shown by name in (a) and the reactions are denoted symbolically in (b).*

TABLE 5.2   Components and Their Symbolic Assignments Based on Wolf *et al.* (2000)

Components	Symbols
Glucose	S_1
Fructose 1,6-bisphosphate	S_2
Hydroxyacetone, glyceraldehydes 3-phosphate, triose-phosphates	S_3
Glycerate 3-phosphate	S_4
Pyruvate	S_5
Acetaldehyde	S_{6ex}
Glycerol	
Ethanol	
ATP	A_3
ADP	A_2
NADH	N_2
NAD	N_1

5.1.1.2   *What Are the Kinetics of the System and Chemical Reactions?*

Notation   With maps and chemical reactions in hand, we can transcribe these reference tables into mathematical symbols. The legend of symbols helps us to easily go back and forth between our noncomputational model descriptions and our developing mathematical description. The legend has been created by assigning symbols to each molecule in our component list (Table 5.2). To maintain a homogenous notation system, we should use the same symbol (S) with an indexed number rather than different symbols for each molecule. This allows us to distinguish readily where in the pathway the substrate exists and can suggest whether the variable is dependent or independent. Dependent variables are assigned symbols first and independent variables are assigned last. Wolf used such a pattern for the substrate backbone of the pathway but broke from the convention when naming the coenzymes (Table 5.3; ATP-A_3, ADP-A_2, NADH-N_2, NAD-N_1). Although the naming pattern used by Wolf seems fairly intuitive, it can become problematic when working with large numbers of coenzymes and effectors or when comparing mathematical models created by different lab groups. We maintain the Wolf notation to make it easy for the reader to go back and forth between this text and the original modeling paper. Once we have assigned symbols for each component, the assigned symbols can be substituted into both our pathway map and chemical reactions (Fig. 5.4; Table 5.3).

Independent variables do not vary during the simulation. Enzymes are taken into account implicitly by the rate equations that characterize the enzymatic activity. In the model here, all enzymes are considered fixed independent variables in that their concentrations do not change. We assume that the rate of change of the total amount of adenonucleotides (A_T) and nicotinamides (N_T) is negligible during the time frame of simulation and therefore the amounts are constant. Although the total amounts of A_T and N_T are independent of time, the amount of ADP versus ATP does change within the time course of the simulation. Thus, A_T and N_T are best understood as independent variables and conserved moieties. In proper biochemical notation, these would be labeled after the other reactions.

$$A_3 + A_2 = A_T$$
$$N_2 + N_1 = N_T$$

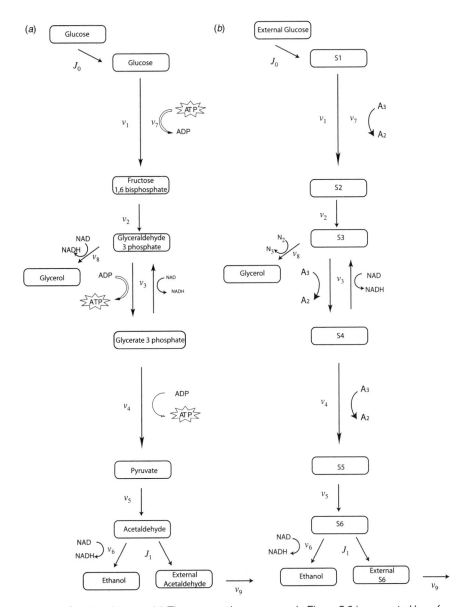

Figure 5.4 *Creating the map. (a) The same pathway as seen in Figure 5.3 is presented here for easy comparison to the completed map. (b) The pathway map shown here substitutes the enzymes with notation for the velocity (V) of the reactions and the influx (J_0) of glucose and efflux (J_1) of acetaldehyde from the cell. Carbon molecules are assigned symbols and numbered sequentially. Cofactors are given separate notation. This is consistent with the notation used by Wolf* et al. *(2000).*

In addition to creating symbolic notation for the components of the reactions, we assign symbols for each biochemical reaction. There are two kinds of notations used to represent rates of change. The first is used for the uptake of glucose. Biologically, this process involves sugar transport across the yeast cell membrane. Although the model does not include physical characteristics, we use notation that indicates the nature of the reaction

TABLE 5.3   Symbolic Reactions

Chemical Process/Reaction	Rate
Cellular influx of glucose	J_0
-> S_1	
Glucose + 2 ATP -> Fructose 1,6-bisphosphate + 2ADP	v_1
$S_1 + 2 A_3$ -> $S_2 + 2 A_2$	
Fructose 1,6-bisphosphate -> (hydroxyacetone, glyceraldehyde 3-phosphate)	v_2
S_2 -> $2 S_3$	
(Hydroxyacetone, glyceraldehyde 3-phosphate) +	v_3
ADP + NAD <-> 3-Phosphoglycerate + ATP + NADH	
$S_3 + A_2 + N_1 = S_4 + A_3 + N_2$	
3-Phosphoglycerate + ADP -> Pyruvate + ATP	v_4
$S_4 + A_2$ -> $S_5 + A_3$	
Pyruvate -> Acetylaldehyde	v_5
S_5 -> S_6	
Acetylaldehyde + NADH -> Ethanol + NAD	v_6
$S_6 + N_2$ -> ethanol + N_1	
ATP -> ADP	v_7
A_3 -> A_2	
(Hydroxyacetone, glyceraldehydes 3-phosphate) + NADH -> Glycerol + NAD	v_8
$S_3 + N_2$ -> Glycerol + N_1	
Acetylaldehyde -> External acetaldehyde	J_1
S_6 -> S_{6e}	
External acetaldehyde -> Degradation or withdrawal	v_9
S_{6e} ->	

in the notation. J is commonly used to identify a flux rate, which is defined generally as flow across an area. The acetaldehyde efflux is also noted with J. The second notation that is dominant in our model and many biochemical reaction models is velocity (v_n) for the rate of mass transiting from one state to another. The number (n) increases in the order of the reaction's appearance within the model.

Create Equations   Writing equations for the model involves at least two steps: (1) the differential equations of the system and (2) writing rate equations for the chemical reactions. Differential equations draw directly from our map and list of reactions with symbolic notation. The equations state how each metabolite changes over time in relation to the other metabolites. The change of any object over time is a rate. We write the rate of change in a metabolite (S) as a relationship between its rates of production and degradation. The rates of production and degradation are indicated subtly in our list of reactions by the arrows between substrate and product and less conspicuously in our flow map where each rate of change is indicated with the symbol v. By writing each equation as, the change in product (dS) over time (dt) is equal ($=$) to the rate of formation (v_f) minus ($-$) the rate of degradation (v_d), we get the list of differential equations shown in Table 5.4.

TABLE 5.4   System Differential Equations

J_0	$\dfrac{dS_5}{dt} = v_4 - v_5$
$\dfrac{dS_1}{dt} = J_0 - v_1$	$\dfrac{dS_6}{dt} = v_5 - v_6 - J_1$
$\dfrac{dS_1}{dt} = J_0 - v_1$	$\dfrac{dS_{6ex}}{dt} = J_1 - v_9^a$
$\dfrac{dS_2}{dt} = v_1 - v_2$	$J_1 = \kappa(S_6 - S_{6ex})$
$\dfrac{dS_3}{dt} = 2v_2 - v_3 - v_8$	$\dfrac{dA_3}{dt} = -2v_1 + v_3 + v_4 - v_7$
$\dfrac{dS_4}{dt} = v_3 - v_4$	$\dfrac{dN_2}{dt} = v_3 - v_6 - v_8$

[a]Wolf *et al.* used a proportionality constant psi (0.1) that is included in the rate constant J_1. κ is a proportionality constant for the efflux of acetaldehyde.

Each v in our set of differential equations maps to a kinetic rate. As the modeler, we define the kinetic rate. It may be a common biochemical or enzymatic rate law. The kinetic rate is not apparent in the notation we use to write the chemical reaction. The modeler chooses the rate law that he or she believes describes the kinetic behavior of the reaction. This decision is based on what the modeler knows of the system and the mediating enzyme. Once chosen, the rate law governs how the reaction behaves in the simulation. We can write each rate (v) as a simple rate equation (law of mass action) in which the reaction rates take the form kS, where S is any of the substrates, $S_1, S_2, \ldots, S_{6e}$. Or we can write more flexible rate equations that can describe the enzymatic rate under a broader set of conditions such as Michaelis-Menten.

Wolf chose to model the glycolytic processes as irreversible except those mediated by glyceraldehyde 3-phosphate dehydrogenase and phosphoglycerate kinase. Under conditions in which reactions are driven in one direction such as near saturating substrate, it is often a reasonable assumption to model the reaction as irreversible. The experimental system moves yeast from glucose-starved to glucose-rich (20 mM) conditions (Richard, 1996). This creates saturating conditions for the influx of glucose, which we model as a constant flux that drives the subsequent reactions forward. Choosing to model the glycolytic reactions as irreversible means that we cannot use this computational model to examine the behavior of glycolysis in situations in which the system is reversible (e.g., high glycerol content and low glucose).

We also assume that mass action kinetics are sufficient to describe the majority of reactions that might otherwise be described by Michaelis-Menten kinetics. For phosphofructokinase, we know that it is a highly regulated enzyme in which there are both activators and inhibitors. However, in this model system, we are focused on the changes in levels of metabolic intermediates and the coenzymes, not the intricacies of phosphofructokinase regulation. Therefore, only inhibition of phosphofructokinase by ATP was included. It was taken into account by including an inhibition constant and cooperativity coefficient (Table 5.5, R2). The general form of the kinetic type used to model phosphofructokinase was Michaelis-Menten (Table 5.5).

5.1.1.3   What Are the Parameters and Initial Conditions for the System?   To
analyze the behaviors of our mathematical model, we must "solve" the differential equations.

TABLE 5.5   Rate Equations for the Chance in Rate Over Time (v)[a]

Rate	Kinetic Function (Rate Equation)	Parameters	Gepasi R No.
J_0	Constant flux	J_0: 50 mM*min^{-1}	R1
V_1	$k_1 S_1 A_3 \left(1/1 + \left(A_3/K_i \right)^n \right)$	k_1: 550 mM^{-1}*min^{-1} K_i: 1.9 mM $n = 4$	R2
V_2	$k_2 S_2$	k_2: 9.8 min^{-1}	R3
V_3	$\dfrac{k_g k_p S_3 N_1 A_2 - k_a k_k S_4 A_3 N_2}{k_a N_2 + k_p A_2}$	k_g: 323.8 mM^{-1}*min^{-1} k_p: 76411.1 mM^{-1}*min^{-1} k_a: 57823.1 mM^{-1}*min^{-1} k_k: 23.7 mM^{-1}*min^{-1}	R4
V_4	$k_4 S_4 A_2$	k_4: 80 mM^{-1}*min^{-1}	R5
V_5	$k_5 S_5$	k_5: 9.7 min^{-1}	R6
V_6	$k_6 S_6 N_2$	k_6: 2000 mM^{-1}*min^{-1}	R7
V_7	$k_7 A_3$	k_7: 28 min^{-1}	R8
V_8	$k_8 S_3 N_2$	k_8: 85.7 mM^{-1}*min^{-1}	R9
J_1	$k(S_6 - S_{6ex})$	k: 375 min^{-1}	R10
V_9	$k_9 S_{6ex}$	k_9: 80.0 min^{-1}	R11

[a]Inhibition of phosphofructokinase by ATP is taken into account by $(A_3/K_i)^n$. When writing these into Gepasi, an asterisk must be included between variables to indicate that they are multiplied (e.g., $k_2 S_2$ is written k2*S2). Spaces are used to distinguish one term from another (e.g., S2 + S3 would be one term, S2 + S3 identifies two). R No. refers to reaction order in Gepasi.

The equations describe our system approximately and can be solved accurately when provided specific values for the parameters (rate constants, concentrations). In other words, we solve for a numerical solution to a single set of initial conditions and parameter values.

Parameter values and initial conditions are chosen that fall within biologically realistic ranges (Table 5.6). The possible ranges are constrained by values obtained in lab experiments. Wolf chose to tune the model parameters around the rate constant (k9) for removal of acetaldehyde from the system. The rate of acetaldehyde removal is determined

TABLE 5.6   Initial Concentrations for Glycolysis Model

	Wolf et al. (2000)
S1: Glucose	1.09
S2: Fructose 1,6-bisphosphate	5.10
S3: Glyceraldehydes 3-phosphate	0.55
A2: ADP	1.71 = A3/A2
A3: ATP	2.19
N2: NADH	0.41
S4: Glycerate 3-phosphate	0.66
S5: Pyruvate	8.31
S6: Acetaldehyde	0.08
N1	N2/N1 = 0.69
External acetaldehyde	0.02
Glycerol	Fixed
Ethanol	Fixed
A2 + A3 =	4
N2 + N1 =	1

experimentally by the rate of KCN or argon addition (Richard *et al.*, 1993; Poulsen *et al.*, 2004). To date, it is understood that the amount of accumulated acetaldehyde must be less than 250 µM for oscillations to occur (Richard *et al.*, 1994, 1996b; Poulsen *et al.*, 2004). Wolf chose the rate constant to be 80 min^{-1} based on a bifurcation analysis that enables modelers to examine the overall behavior of the set of equations. The range of NADH concentrations possible with this value of k9 were within the range seen by Richard.

5.2 COMPUTATIONAL MODEL

It is worth pausing at this time to review the steps we have completed. We have used topological maps (diagrams of metabolic pathways); written explicit chemical reactions; assigned symbols to each component; created differential equations that describe the behavior of the pathways (symbolically); and chosen rate laws to describe the enzymatic behaviors of each reaction (law of mass action and others defined by Wolf). Each step was performed without using a computer, and yet we have created a mathematical model of the biological process. Now that we have characterized all features of the model, we can use a numerical solver to simulate the behavior of the model.

Automated solvers are available in software applications for general and specialized topics. General solvers include such software as MATLAB and Mathematica, which are commonly used in math and engineering disciplines. The ODE table (Table 5.5) would be used to create the mathematical model in these tools. In the tool we will work with, Gepasi, we enter the reactions (the components and their relationships), rate equations, and parameters that describe the model. These features are then used by the biochemical simulator to construct the ODEs and algebraic equations that are solved for the simulation.

5.2.1 Software: Gepasi

Gepasi (Mendes, 1993) is one of a series of tools known as a general simulator; others include SCAMP, METAMODEL, and MIST (Cornish-Bowden and Hofmeyr, 1991; Sauro, 1991, 1993; Ehlde and Zacchi, 1995). These tools for modeling biochemical processes vary in their design, user interfaces, and range of utility. Gepasi has a graphical user interface that allows researchers to enter reactions with standard biochemical notation rather than a programming language (Mendes, 1993; Sauro, 1993). Tools for modeling, some more than others, often require training sessions for users to become fully fluent with them. Choosing a simulator for your research group is based on your comfort using the interface to enter information, navigate and view results, as well as the package's overall functionality. A number of biochemical kinetic simulators can be downloaded and tried for free. These software are tailored for biologists and biochemists.

We are using Gepasi because it has a friendly user interface, provides a set of kinetic rate laws to choose from, and is free for nonprofit use. The method for entering the model reactions is similar to writing biochemical reactions of the biological process. As such it maps well to early lessons in writing equations for biological processes and can be used at multiple academic levels. As a research tool, Gepasi is significant in terms of its computational capabilities. It is one of the first simulators to have a well-developed parameter estimation tool. It also supports metabolic control analysis and parameter scanning.

To use Gepasi, we must first go to the Gepasi Web site (http://www.gepasi.org), download the software from the site to a local drive. The downloaded program installs

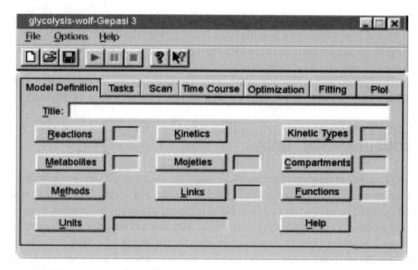

Figure 5.5 *Gepasi Model Definition window. The Model Definition window is the first to appear when Gepasi is started. Gepasi has a Windows-like interface such that at the top of the frame there are commands in the menu dialogues "File," "Options," and "Help," and an icon toolbar. The "Model Definition" tab is selected, the left foremost tab. Title and comments are written in the text fields provided in this window. Buttons are used to navigate through elements of the model that are to be defined by the user.*

Gepasi. Installation directions are provided at the site. The software is compatible with the following operating systems: Windows 95/98/2000/NT and XP. A tutorial designed to introduce new users to Gepasi is also provided at the Web site (see tutorial on signal transducers).

5.2.1.1 Entering the Model: Model Definition

Each software tool has a different interface for entering a model. This is why it is helpful to have developed a biological system model prior to using any given simulation tool. This makes adjusting the software independent of defining one's model. When we open Gepasi, the initial window that appears is "Model Definition" (Fig. 5.5). This is where the features of the model, reactions, and kinetics, are entered. Each model is given a specific name, "title." The Gepasi model we are discussing now will be referred to as Glycolysis-Wolf.

Reactions   The noncomputational model of our metabolic pathway contains the reactions that we will simulate (Table 5.3). Within Gepasi, multiple reactions can be entered by clicking on the "Reactions" button and writing out each reaction formula. Reactions are assigned a reaction "Name," that is, R1, R2, according to the order in which they are typed into the software. You can also enter your own name for the reaction manually. The "Help" button provided gives instructions on the syntax for writing the chemical reaction.

There are two main considerations when writing the equation in Gepasi syntax. One is the structure of the equation and the other is the content. In Figure 5.6, we can see the syntax for the first two reactions of our system, the influx of glucose (S1) and the conversion of glucose to glucose 6-phosphate (S2). The syntax uses combinations of letters, numbers, and symbols to describe the variables and directionality of the reaction. The letter S and number 1 are recognized as a single variable named "S1." The hyphen (-) and greater than (>) symbols together indicate a forward irreversible reaction. Spaces are critical for the appropriate recognition of variables or directions. For example, placing a

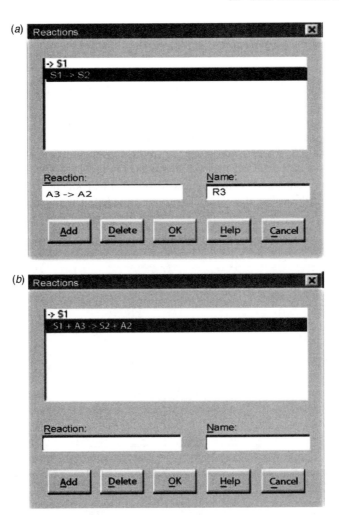

Figure 5.6 *Gepasi Reaction Window. Reactions are written by the user into the bottom left field "Reaction." Gepasi sequentially assigns a "Name" to each reaction as it is entered. (a) Two reactions have been entered, the influx of glucose, S1, and conversion of glucose to glucose 6-phosphate, S2. A separate reaction for the ATP, A3, to ADP, A2, could be considered. (b) Reaction 2, R2, has been written to include the cofactor ATP, A3. The second reaction of (a) and (b) represent two different types of reactions.*

space between the letter "S" and number "1" to create "S 1" instead of "S1" creates two variables: one named "S" the other named "1." The same combination without the space creates one variable named "S1."

We next consider the equation content. Reaction 2, the conversion of glucose to glucose 6-phosphate, involves the hydrolysis of 2ATP to 2ADP (Table 5.3; Fig. 5.6b). It might be tempting to identify the hydrolysis as a separate reaction, creating two simple reactions of S1 -> S2 and A3 -> A2 (Fig. 5.6a). However, the kinetics of the conversion of S1-S2 is dependent on ATP hydrolysis. This is taken into account in the rate equation. Figure 5.6b shows the same two reactions as Figure 5.6a. However, in Figure 5.6b the second equation is written with the cofactors ATP (A_3) and ADP (A_2) included. The content and syntax of the equation determines the type of kinetics Gepasi suggests for the reaction in later steps.

Use the list of reactions in Table 5.3 to enter the remaining equations into Gepasi. When done, one should have 11 reactions.

Kinetics   To set the kinetics for the reactions we entered in Gepasi, click on the "Kinetics" tab from the main window. The reactions we created are shown in the left frame and a list of kinetic types on the right (Fig. 5.7). Gepasi presents a selection of kinetic types based on the syntax of the highlighted reaction. The first three reactions of our model (R1, R2, R3) are irreversible reactions as indicated by our use of the unidirectional arrow (->). However, the kinetic options presented when each is highlighted differ. This is because of differences in the number of substrates and products involved. Each kinetic type chosen for the reaction requires the value of a parameter to be provided. The parameter values enable

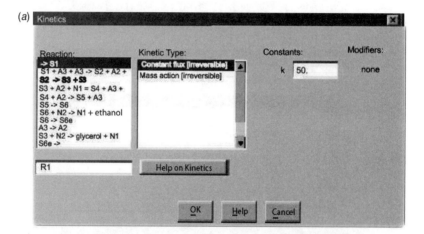

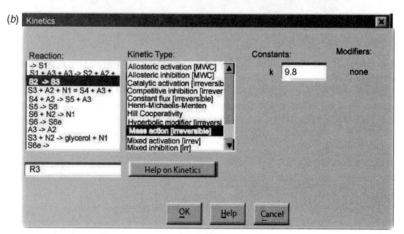

Figure 5.7   *Gepasi Kinetics Window. (a, b) Images of the Kinetics window in Gepasi. The left "Reaction" panel contains the set of reactions entered for the model. When a reaction is highlighted, its reaction name is shown in the lower left field (a, R1; b, R3). The right-hand "Kinetic Type" panel lists kinetic rate equations based on the syntax of the highlighted reaction. When a kinetic type is selected, the constants or modifiers required for the rate equation are indicated to the top right corner of the frame. (a) Constant flux is selected to model the influx of glucose (S1) in the first reaction (R1) of the model. (b) Mass action (irreversible) has been selected for the conversion of glucose 6-phosphate (S2) to the triose phosphates (S3).*

Gepasi to solve the ordinary differential equations associated with the reaction. The values we enter are ideally based on experimental findings.

The kinetics of each reaction are indicated in Table 5.5. Reaction R1 is written as a flux with no substrate specified, "constant flux (irreversible)." The parameter value for the irreversible influx of glucose is written as a velocity (J). Reaction 2 has substrates that react at some rate that determines the velocity at which glucose (S1) is converted to (S2), "Mass Action (irreversible)." The parameter value for the rate at which the substrates react is indicated as a rate constant (k). Reaction 3 is a unidirectional, one substrate, one product reaction. Gepasi presents a series of predefined kinetic rates that may be used to model the reaction. The "Help on Kinetics" button shows rate equations for each named rate law.

There will be times when one says, "None of the rate laws that I expected are showing up as a choice." This is a time to double-check your syntax. Again, improper spacing can lead to two substrates appearing as one. For example, you might consider writing R2 as $S_1 + 2 A_3 -> S_2 + 2 A_2$. Note the spacing of numbers and letters. Gepasi requires us to rewrite the equation to $S_1 + A_3 + A_3 -> S_2 + A_2 + A_2$. Otherwise, instead of creating two molecules of A_3 or A_2, we would create new metabolites named 2, A_3 and A_2.

Another reason for not seeing an appropriate rate law is that you need to define a new one. When the enzymatic mechanism does not fit previously defined kinetic rate laws, you can click on the "Kinetic Types" button to write your own equation to describe it. We need to do this for reaction 2 (R2) in order to take into account inhibition of phosphofructo-kinase by ATP and for reaction 4 (R4), whose reversible kinetics involves the rate constants for two enzymes. The details of creating a kinetic type can be found in a tutorial on the Gepasi Web site [Mendes, 1997; http://www.gepasi.org (see tutorial on signal transducers)]. For any newly created rate equations, you must indicate the number of substrates and products associated with the kinetic model (i.e., two substrates, one product).

Metabolites When all the reactions of the Wolf model are entered, the number 13 appears next to the "Metabolites" button. The initial concentrations of metabolites are entered in the "Metabolites" window. Gepasi identifies the metabolites from the entered reactions. "Metabolites" is where components of each reaction are designated as fixed or variable. A check box is provided to indicate when a concentration is fixed (constant) throughout the simulation, making it a parameter or boundary conditions. The default setting is for components in the reactions to be considered variables. A default concentration of 0.00001 mM, written in Gepasi as 1.e-005, is given to each metabolite. The initial concentrations we use are the mean concentration values Wolf used (Table 5.6; Fig. 5.8). Glycerol and ethanol are set as fixed values of 1. Why? These two end products are produced but not monitored in our system. These types of fixed variables are called boundary conditions or sinks. As fixed concentrations, they do not affect the rate of any other reaction but they do provide a sink for mass of the system.

Reactions and their associated metabolites are created within a single compartment. Compartments create boundaries for reactions and their metabolites. Given that Gepasi was developed as a 2D simulator, spatial dimensions are not taken into account in the simulations, only the reactions and time. Compartments act as a generic identifier of reactions and metabolites that are available to one another. By creating more than one compartment, metabolites, and reactions can be separated from one another. Gepasi allows for compartments to be linked together simulating transfer across membranes and allowing users to set different conditions (parameters) in each compartment and comparing

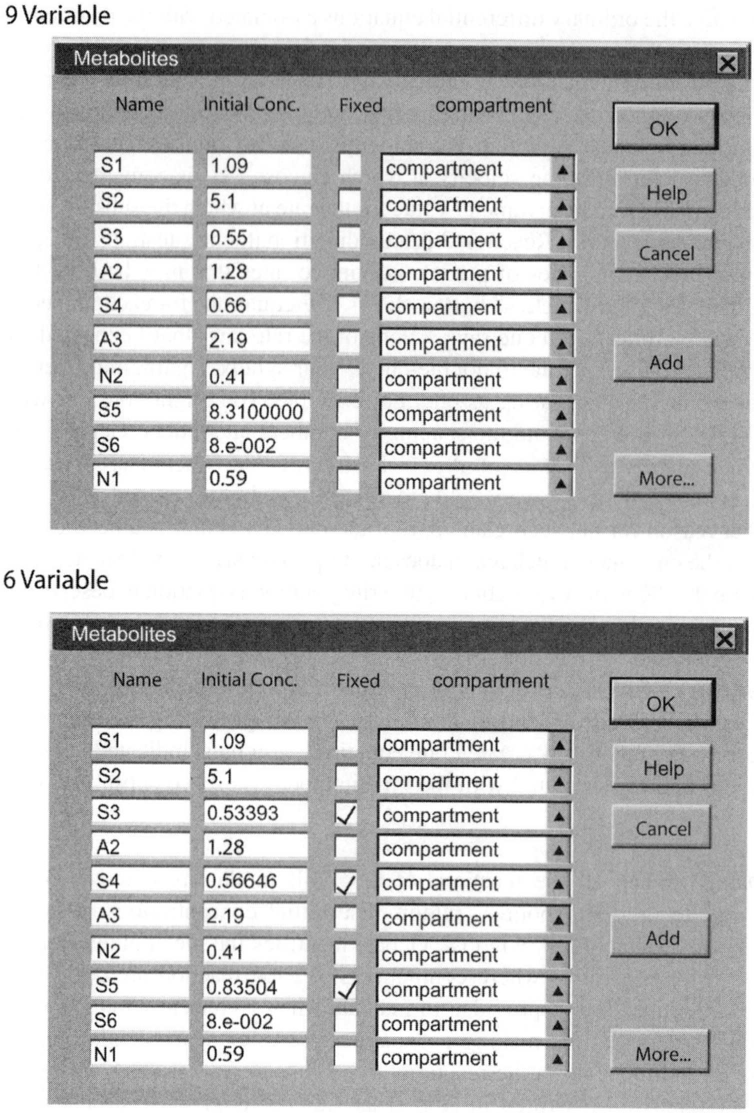

9 Variable

Metabolites				
Name	Initial Conc.	Fixed	compartment	OK
S1	1.09	☐	compartment ▲	
S2	5.1	☐	compartment ▲	Help
S3	0.55	☐	compartment ▲	
A2	1.28	☐	compartment ▲	Cancel
S4	0.66	☐	compartment ▲	
A3	2.19	☐	compartment ▲	
N2	0.41	☐	compartment ▲	Add
S5	8.3100000	☐	compartment ▲	
S6	8.e-002	☐	compartment ▲	
N1	0.59	☐	compartment ▲	More...

6 Variable

Metabolites				
Name	Initial Conc.	Fixed	compartment	OK
S1	1.09	☐	compartment ▲	
S2	5.1	☐	compartment ▲	Help
S3	0.53393	☑	compartment ▲	
A2	1.28	☐	compartment ▲	Cancel
S4	0.56646	☑	compartment ▲	
A3	2.19	☐	compartment ▲	
N2	0.41	☐	compartment ▲	Add
S5	0.83504	☑	compartment ▲	
S6	8.e-002	☐	compartment ▲	
N1	0.59	☐	compartment ▲	More...

Figure 5.8 Gepasi Metabolites window. Initial concentrations for the model variables are entered. The check box allows us to fix metabolite concentrations to test the hypothesis of this model. Lower panel shows Wolf values for six-variable simulation. Gepasi may reexpress values in scientific notation as can be seen for S6, which was entered as 0.08. The "More ... " button must be clicked to access the remaining variables and enter all initial values.

results. This creates a series of 2D simulations that use the results of one compartment as parameters or initial conditions in another.

Moieties Gepasi recognizes mass conservation within the set of reactions entered. This means that Gepasi generates a set of algebraic equations (i.e., $A_2 + A_3 = A_T$ and $N_1 + N_2 = N_T$). These equations tell the computer that the total amounts of adenine and nicotinamidic molecules are independent of time even as the amount of any one molecular state (A_2, A_3, N_1, N_2) varies dependently over time. The total mass of the conserved variables is calculated from the concentrations entered from the metabolites.

Methods   In order to run any model as a simulation, we must choose the method for solving the set of differential equations that describe how the system behaves. In general, the ordinary differential equations in Gepasi are invisible to users. The kinetic rate laws are the only glimpse of equations. They are a small part of the overall set of equations that define the model mathematically. The specific parameter values we provide allow us to solve numerically our set of system equations. Therefore, we must choose a numerical method.

The "Method" button opens a panel in which we choose one of four numerical methods for solving the ordinary differential equations of the system. Gepasi uses the set of solvers known as Livermore Solver Of Differential Equations (LSODA) (Hindmarsh, 1983; Mendes, 1993; Petzold, 1983). LSODA automatically selects one of two approximation methods to use during the simulation based on the stiffness of the equations. Stiffness is a function of the number of scales within the differential equations. For example, in our model, the rate constant for the production of pyruvate is essentially 9 mM/min, whereas the rate constant for glycerol kinase is 2000 mM/min. When equations are non-stiff, the Adams method is used. For stiff regions, the Backward Differentiation Formulas (BDF) method is used. LSODA is sufficiently robust for solving differential equations that if the simulation fails, it is likely due to a faulty equation or relationship in the reactions. The "derivation factor" is a feature of numerical solutions that sets how big or small the finite step from one value to the next is in the calculation of a derivative. The value is written as a percentage (i.e., if 0.1 is entered, it is read as 0.1%). The default settings are sufficient for the simulations we wish to perform.

The methods available for steady-state analysis include Newton, integration, Newtonian and integration, or backward integration (Fig. 5.9). The default setting for the computational method for finding the steady state is Newton+integration with back integration if all else fails (Fig. 5.9). Steady-state resolution (S.S. resolution) sets the value for the amount of change that can occur within the variables and still have the system considered to be at a steady state. By setting S.S. resolution to 0.00005 (5×10^{-5} or 50 nm in a mM unit simulation), the concentrations must have a change greater than this amount to no longer be at steady state. Gepasi Help pages provide additional information on other settings and their mathematical formulations.

5.2.2 Simulation Controls

Each simulation tool has to be told by the user what to do with the model that has been defined. To enable the user to define the simulation steps, Gepasi has created a series of menu tabs: "Task," "Scan," "Time Course," "Optimization," "Fitting," and "Plot." We will focus on an overview of the tabs required to run an initial simulation of the modeled system: Task and Plot.

5.2.2.1 Tasks   In this window, we define the type of experiment we are running in the system, a time-course study or steady-state analysis (Fig. 5.10). For a time-course study, we select the "end time" of the reaction and the number of time points to be reported within the time course. The "Edit" button is used to select the parameters that will be monitored during the simulation. The selected metabolites are then shown in the window to the right. The collected data is saved to a default ASCII file named "tcsimresults.dyn" or

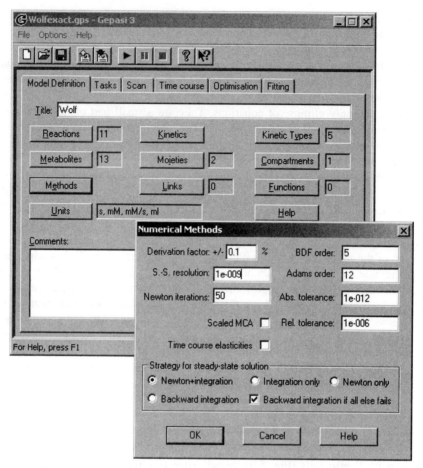

Figure 5.9 *Gepasi Methods window. "Methods" button from the main GUI window opens the Numerical Methods window. Dialogue boxes are used to set the parameters to approximate the solution to the ordinary differential equations developed by Gepasi based on the reactions and kinetics of the model. The default settings for BDF and Adams order are sufficient for this simulation.*

"simresults.ss." The type of simulation is selected by checking off the check box directly under "Time Course" or "Steady State."

5.2.2.2 Plot Gepasi files are graphed with third-party software, Gnuplot. Gnuplot imports and displays the data stored in the simresult.dyn file (Fig. 5.11). The same ASCII file may be imported to Excel or other graphing software. By examining the ASCII files, we can see each data point value collected during the simulation.

5.3 RESULTS

5.3.1 Nine-Variable Model

Richard found experimentally that sustained macroscopic oscillations in glycolysis were accompanied by oscillations in the six-carbon (C_6) sugars and that the three-carbon

(a)
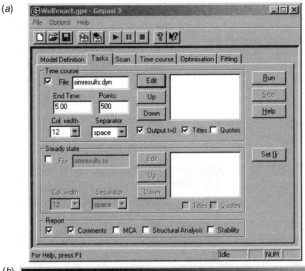

(b)
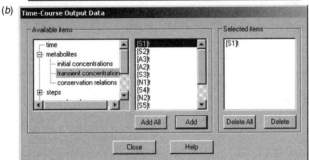

(c)
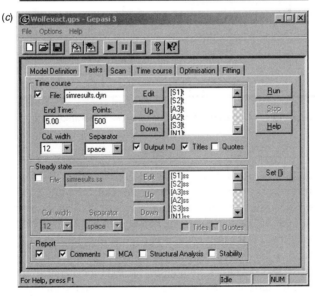

Figure 5.10 *Gepasi Tasks window. (a) The Task window is used to tell the software how long to run the simulation ("End Time"), how many time points to collect ("Point"), and where to save the simulation results ("File"). (b) Variables and fluxes must be selected to be monitored during the simulation. They are selected using the "Edit" button. (c) Monitored variables are visible in the Tasks window.*

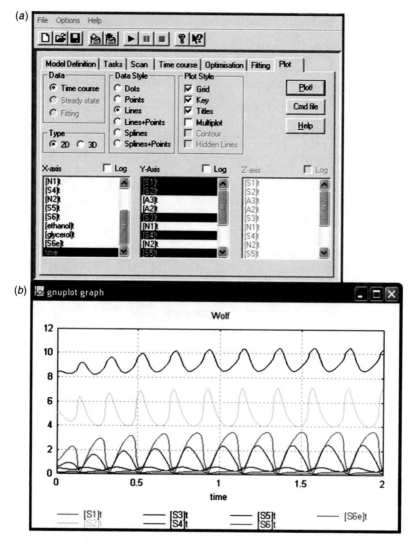

Figure 5.11 *Screenshots of Gepasi plot tab (a) and Gnuplot generated graph of simulation results (b). Variables highlighted in the X-axis and Y-axis columns (a) are plotted accordingly in the Gnuplot graph.*

sugar molecules (S3, S4, S5) did not oscillate. Richard argued that the relative amplitude of the driving reactions should be greater than the relative amplitude of the driven oscillations. Based on the observed experimental results, oscillations along the sugar backbone were insufficient to drive oscillations in the subsequent metabolites (Fig. 1 from Richard). Thus, he proposed that ATP/ADP oscillations are sufficient to drive NADH oscillations when no oscillations are present in downstream metabolites. The Wolf model, which we have re-created, was developed to address the hypothesis put forth by Richard.

The model that we have created is designed to examine the plausibility of a molecular mechanism based on what we already know about yeast enzyme kinetics in anaerobic conditions and with glycolytic oscillations induced. The modelers were not conducting experiments beyond those performed by Richard and therefore were not developing additional constraints that could be used to tune the parameters of the model to experiental values.

The qualitative behavior of the model is what we use to assess whether or not the model sufficiently reproduces the phenomenon of interest. With the model implemented in Gepasi, we can begin to examine the behavior of the model. The nine-variable system of Wolf with parameter values from Table 5.6 mimics behaviors of the yeast anaerobic system of Richard in the following ways:

Oscillations are found throughout the glycolytic chain and in the energy carriers (Fig. 5.12).

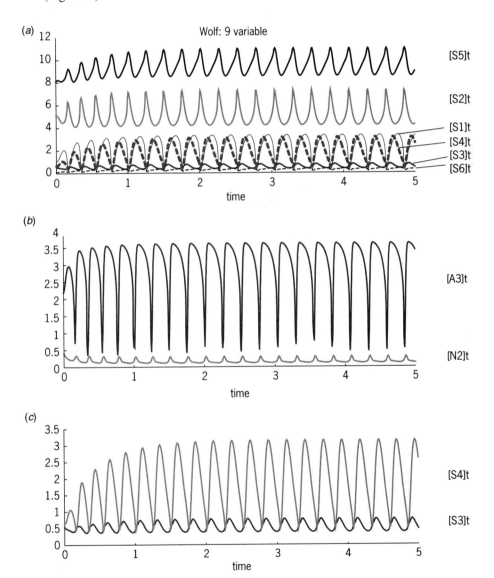

Figure 5.12 *Results of nine-variable simulation. The simulation results are plotted using GnuPlot. Persistent oscillations develop in all six variables of the pathway backbone (graph a: S1–S6) and the energy carriers (graph b: A3, N2). S4 oscillations are significantly higher than the precursor metabolite S3 (graph c). The nine-variable model was simulated with parameter values from Table 5.5 and initial values from Table 5.6. The numerical method parameters were default values for derivation factor (0.1), BDF (5), Adams order (12), and default relative tolerance (0.000001).*

TABLE 5.7   Relative Amplitudes of Select *In Vivo* and Simulated Variables

System	S3:S4	ATP/ADP:S4	ATP/ADP:NADH/NAD
Richard[a]	34%:73%	98%:73%	98%:53%
Nine variable	0%:0%	75%:NF	75%:39%
Six variable	0%:0%	71%:0%	71%:15%

[a]Modified results from Richard *et al.* (1996); NF, no fit to sinus curve.

> The relative amplitude in S3 oscillations is smaller than the relative amplitude in S4 (Fig. 5.12, Table 5.7).

> The relative amplitude of oscillations in the ration of ATP/ADP is greater than S4 (Table 5.7).

Given the model behaves as the experimental system in the behaviors that we wish to examine (Fig. 5.12), we can next ask the question: do oscillations within the six-carbon sugars and NADH (N2) still occur when S3, S4, S5 are constant? Unlike the experimental system in which fixing the concentrations to a constant value is not possible, in our computational model, we can eliminate the possibility of oscillations occurring by setting them to fixed values. Once the metabolite concentrations are fixed, any subsequent oscillations in the unfixed metabolites must be attributed to other factors in the model.

5.3.2   Fixed Variables: Six-Variable Model

Fixing the concentrations of variables redefines the set of equations that describe the biology of the system. The differential equations describing the rates of change are set to equal zero and the value of the variable is held constant. Mass still moves through the system, but it moves such that the change in concentration over time for the fixed variable is zero. Wolf chose to fix the concentrations of S3, S4, and S5 at values that were biologically realistic and that satisfied the mathematical constraint of zero change in concentration over the oscillation period. Under these conditions, using the values identified by Wolf, we see that oscillations persist in the six-carbon sugars S1, S2 and the energy carriers A3, N2 (Fig. 5.13, Table 5.7). Oscillations persist in NADH in the absence of oscillation within the three-carbon sugars, which is consistent with the observations and hypothesis of Richard.

5.4   CONCLUSION

The strength of modeling this system is that it enabled the researcher, in this case Wolf, to explore the prediction of Richard that oscillations in the six-carbon sugars and energy carriers were sufficient to propagate oscillations through the pathway. To achieve this experimentally, the enzymes producing the three-carbon sugars would have to be simultaneously inhibited without damaging the biological system and creating additional artifacts. The model enabled Wolf to perform an experiment that was otherwise unfeasible in the *in vivo* system. With the model in hand, you can explore other scenarios for metabolite concentrations or enzymes functioning at different rate constants. For example, we can ask additional questions of this model. For instance, the amplitude in NADH

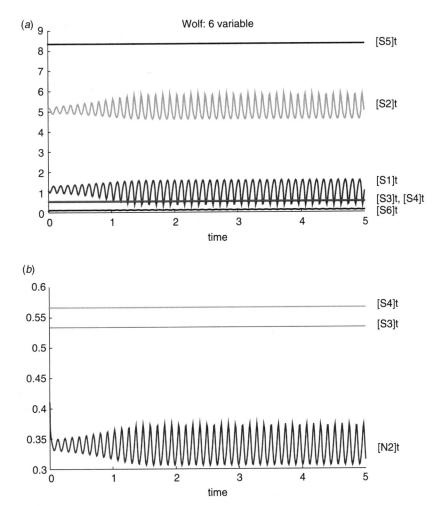

Figure 5.13 *Six-variable simulation. GnuPlot of six variables from simulation in which S3, S4, and S5 were fixed at 0.53393, 0.56646, and 0.83504, respectively. Oscillations develop and persist in S1 and S2 (a) and N2 (b).*

oscillations in the six-variable model appears to be less than that of the nine-variable model. We can change initial conditions (starting concentrations) and parameters (rate constants) to develop a better understanding of how the factors in the six-variable model affect the phase, frequency, and amplitude of NADH oscillations. Wolf went on to use the model to examine acetaldehyde as a synchronization mechanism for the production of macroscopic population oscillations.

We chose the Wolf model because it clearly drew on experimental data and experiments, highlighted the use of models to explore the plausibility of a mechanism, and retained sufficient biological detail to be recognizable and believable as a model of glycolysis. Other published glycolysis models have been made about different model organisms and nonoscillating conditions (Richard, 2003). Many of these models are accessible in online repositories in Systems Biology Markup Language (SBML) or as java applets online (JWS) (Bier *et al.*, 1996; Teusink *et al.*, 1996; Hucka *et al.*, 2003; Olivier and

Snoep, 2004). Previously published models and model repositories are valuable for learning more about modeling glycolysis. The models contain the pathway structure, present the equations used to model the reaction, and allow the user to change the parameter values used by the system. Between publicly available models, software, and research articles, it is possible from an educational perspective to construct resources that allow novices to better understand the methods, benefits, and shortcomings of modeling glycolysis.

BIBLIOGRAPHY

Aon MA, Cortassa S, Westerhoff HV, *et al.* (1992). Synchrony and mutual stimulation of yeast cells during fast glycolytic oscillations. *Journal of General Microbiology* 138:2219–2227.

Betz A, Chance B (1965). Influence of inhibitors and temperature on the oscillations of reduced pyridine nucleotides in yeast cells. *Archives of Biochemistry and Biophysics* 109:579–584.

Betz A, Hinrichs R (1968). Incorporation of glucose into an insoluble polyglycoside during oscillatory controlled glycosis in yeast cells. *European Journal of Biochemistry* 5:154–157.

Bier M, Teusink B, Kholodenko BN, *et al.* (1996). Control analysis of glycolytic oscillations. *Biophysical Chemistry* 62:15–24.

Bier M, Bakker BM, Westerhoff HV (2000). How yeast cells synchronize their glycolytic oscillations—a perturbation analytic treatment. *Biophysical Journal* 78:1166–1175.

Chance B, Estabrook RW, Ghosh A (1964a). Damped sinusoidal oscillations of cytoplasmic reduced pyridine nucleotide in yeast cells. *Proceedings of the National Academy of Sciences USA* 51:1244–1251.

Chance B, Schoener B, Elsaesser S (1964b). Control of the waveform of oscillations of the reduced pyridine nucleotide level in a cell free extract. *Proceedings of the National Academy of Sciences USA* 52:337–341.

Cornish-Bowden A, Hofmeyr JHS (1991). MetaModel: a program for modelling and control analysis of metabolic pathways on the IBM PC and compatibles. *Computer Applications in the Biosciences* 7:89–93.

Das J, Busse HG (1985). Long term oscillation in glycolysis. *Journal of Biochemistry* 97:719–727.

Ehlde M, Zacchi G (1995). MIST: a user-friendly metabolic simulator. *Computer Applications in the Biosciences* 11:201–207.

Ghosh A, Chance B (1964). Oscillations of glycolytic intermediates in yeast cells. *Biochemical and Biophysical Research Communications* 16:174–181.

Goldbeter A, Lefever R (1972). Dissipative structures for an allosteric model. Application to glycolytic oscillations. *Biophysical Journal* 12:1302–1315.

Heinrich R, Rapoport TA (1975). Mathematical analysis of multienzyme systems. II. Steady state and transient control. *Biosystems* 7:130–136.

Hess B, Boiteux A (1968). Mechanism of glycolytic oscillation in yeast. I. Aerobic and anaerobic growth conditions for obtaining glycolytic oscillation. *Hoppe Seylers Z Physiology Chemistry* 349:1567–1574.

Hindmarsh AC (1983). ODEPACK, A Systematised Collection of ODE Solvers. In: *Scientific Computing*. In: Stepleman RS *et al.*, eds. North-Holland: Amsterdam; pp. 55–64.

Hucka M, Finney A, Sauro HM, *et al.* (2003). The systems biology markup language (SBML): a medium for respresentation and exchange of biochemical network models. *Bioinformatics* 19(4):524–531.

Hynne R, Dano S, Sorensen PG (2001). Full-scale model of glycolysis in Saccharomyces cerevisiae. *Biophysics and Chemistry* 94:121–163.

Kanehisa M (1997). A database for post-genome analysis. *Trends in Genetics* 13:375–376.

Kanehisa M, Goto S (2000). KEGG: Kyoto Encyclopedia of Genes and Genomes. *Nucleic Acids Research* 28:27–30.

Kanehisa M, Goto S, Hattori M, *et al.* (2006). From genomics to chemical genomics: new developments in KEGG. *Nucleic Acids Research* 34:D354–D357.

de Koning W, van Dam K (1992). A method for the determination of changes of glycolytic metabolites in yeast on a subsecond time scale using extraction at neutral pH. *Analytical Biochemistry* 204:118–123.

Mendes P (1993). GEPASI: A software package for modeling the dynamics, steady states and control of biochemical and other systems. *Computer Applications in Biosciences* 9:563–571.

Mendes P (1997). Biochemistry by numbers: simulation of biochemical pathways with Gepasi 3. *Trends in Biochemical Sciences* 22:361–363.

Ogata H, Goto S, Sato K, *et al.* (1999). KEGG: Kyoto Encyclopedia of Genes and Genomes. *Nucleic Acids Research* 27:29–34.

Olivier BG, Snoep JL (2004). Web-based kinetic modelling using JWS Online. *Bioinformatics* 20(13):2143–2144.

Petzold LR (1983). Automatic selection of methods for solving stiff and nonstiff systems of ordinary differential equations. *SIAM Journal on Scientific and Statistical Computing* 4:36–148.

Poulsen AK, Lauritsen FR, Olsen LF (2004). Sustained glycolytic oscillations-no need for cyanide. *FEMS Microbiology Letters* 236:261–266.

Reijenga KA, Snoep JL, Diderich JA, *et al.* (2001). Control of glycolytic dynamics by hexose transport in Saccharomyces cerevisiae. *Biophysical Journal* 80:626–634.

Richard P (2003). Rhythm of yeast. *FEMS Microbiology Review* 27:547–557.

Richard P, Teusink B, Westerhoff HV, *et al.* (1993). Around the growth phase transition *S. cervisiae*'s make-up favours sustained oscillations of intracellular metabolites. *FEBS Letters* 318:80–82.

Richard P, Diderich J, Bakker B, *et al.* (1994). Yeast cells with a specific cellular make-up and an environment that removes acetaldehyde are prone to sustained glycolytic oscillations. *FEBS Letters* 341:223–223.

Richard P, Teusink B, Hemker MB, *et al.* (1996a). Sustained oscillations in free-energy state and hexose phosphates in yeast. *Yeast* 12:731–740.

Richard P, Bakker BM, Teusink B, *et al.* (1996b). Acetaldehyde mediates the synchronization of sustained glycolytic oscillations in populations of yeast cells. *European Journal of Biochemistry* 235:238–241.

Richter O (1974). Computer simulation der Glycolyseoszillationen: Ein Vergleich zwischen Modellen und Experimenten [PhD Thesis]. Mathematisch Naturwissenschaftliche Fakultat, Fridich-Wilhems Universitat, Bonn.

Richter O, Betz A, Giersch C (1975). The response of oscillating glycolysis to perturbations in the NADH/NAD system: a comparison betweens experiments and a computer model. *BioSystems* 7:137–146.

Sauro HM, Fell DA (1991). SCAMP: A metabolic simulator and control analysis program. *Mathematical and Computer Modelling* 15:15–28.

Sauro HM (1993). SCAMP: a general-purpose simulator and metabolic control analysis program. *Computer Applications in the Biosciences* 9(4):441–450.

Teusink B, Larsson C, Diderich J, *et al.* (1996). Synchronized heat flux oscillations in yeast cell populations. *Journal of Biological Chemistry* 271:24442–24448.

Teusink B, Westerhoff H (2000). 'Slave' metabolites and enzymes. *FEBS Journal* 267: 1889–1893.

Voit E, Ferreira A (2000). *Computational Analysis of Biochemical Systems: A Practical Guide for Biochemists and Molecular Biologists.* Cambridge: Cambridge University Press; 544p.

Wolf J, Passarge J, Somsen OJG, *et al.* (2000). Transduction of intracellular and intercellular dynamics in yeast glycolytic oscillations. *Biophysics Journal* 78:1145–1153.

Wolf J, Sohn H, Heinrich R, *et al.* (2001). Mathematical analysis of a mechanism for autonomous metabolic oscillations in continuous culture of *Saccharomyces cerevisiae*. *FEBS Letters* 499:230–234.

Yuan Z, Medina MA, Boiteux A, *et al.* (1990). The role of fructose 2,6-biphosphate in glycolytic oscillations in extracts and cells of *Saccharomyces cerevisiae*. *European Journal of Biochemistry* 192:791–795.

Chapter *6*

Cell Cycle

The cell cycle is a beautifully regulated system and is studied intensely because of its direct relationship to development and growth, whether that is in the arrested cell cycles of gametes, embryonic development through rapid divisions, or cancerous growth due to abnormally regulated cells. Cell cycle phases impact cellular structures and functions. Protein synthesis is inhibited, cytoskeletal structures are reorganized, and DNA is replicated and segregated. It may be helpful to take a moment to clarify that cell cycle and cell growth, although intimately related, are not the same. The cell cycle is composed of the protein dynamics that enable cell growth. When the cell cycle is too short because of faulty molecular interactions of the cellular machinery, cells fail to increase sufficiently in size. In turn, if the cell cycle is unable to complete the process of mitosis, no cell division occurs, and the population of cells does not grow. *Cell growth* is a macroscopic property that refers at the individual cell level to the size of the cell, but *cell growth* when applied to populations refers to the number of cells in the population. Throughout this chapter, we will use *cell cycle* to refer to the molecular interactions that drive the process of cell division.

Biochemical and cellular imaging studies in *Xenopus laevis*, starfish, sea urchins, and mice have tightly correlated detailed information on gene and protein expression patterns with cell cycle progression and inhibition. The wealth of knowledge concerning the molecular patterns and functions of the cell cycle proteins makes the cell cycle an exciting system to model. The difficulty in modeling the system, however, is that little is known quantitatively about the kinetic rates of the individual reactions and bindings.

In this chapter, we will work with a minimal model of the cell cycle engine. These models contain the fewest variables and parameters necessary for reproducing embryonic mitotic cell cycle behaviors. Minimal models reflect the initial understandings of the biological factors involved. The molecular details of cyclin and cyclin-dependent kinase

A Cell Biologist's Guide to Modeling and Bioinformatics. By Raquell M. Holmes
Copyright © 2007 John Wiley & Sons, Inc.

interactions that are the molecular machinery are indirectly reflected in the gross behavior of the system. More detailed models of the cell cycle engine have also been developed. The number of variables and parameters in these more detailed models makes them unwieldy for the purpose of this textbook. However, these models, particularly those developed by Novak and Tyson, are also of great interest because of the experimentation done in parallel, which highlights the usefulness of models in producing testable hypotheses.

6.1 CELL CYCLE CHARACTERISTICS

6.1.1 Protein Patterns

The cell cycle engine has been studied and reviewed in great detail in other texts. We will focus in the following sections on highlights of the cell cycle relevant to the development of the model discussed and refer readers to other materials for more in-depth discussions (Doree, 1990; Nurse, 1990; Nugent *et al.*, 1991; Murray and Hunt, 1993). It is helpful to begin with an understanding of the biological features that we want to examine with our model. The patterns generated by cell cycle proteins are briefly described below.

The driver of the cell cycle is MPF. MPF was identified through functional assays in which oocytes arrested in meiosis were induced to obtain metaphase characteristics by the injection of cytoplasm from other metaphase staged oocytes (Fig. 6.1) (Masui and Markert, 1971). Once its role in mitosis was also recognized, the factor originally dubbed maturation promoting factor (MPF) became known as M-phase promoting factor. The characteristic behavior of MPF is that its activity is low during interphase and high during meiosis or mitosis (i.e., M-phase). It was later determined that MPF is a heterodimer of two proteins, cyclin and a cyclin-dependent kinase (cdc2) (Labbe *et al.*, 1988, 1989; Gautier *et al.*, 1990). The kinase is regulated by binding to cyclin and subsequent phosphorylation states. The schematic provided in Figure 6.2 maps out the series of core reactions that take place to drive the cell cycle forward. We will review the characteristics of the major components of the cell cycle in the following paragraphs and refer to mitotic cyclins as cyclin and to cyclin-dependent kinases as cdc2 kinase.

Cyclins are a family of proteins whose expression increases and decreases in relation to cell cycle stages. Numerous protein gels and Western blots have shown that mitotic cyclins are synthesized throughout interphase: their levels reach and maintain a peak during M-phase and drop precipitously at anaphase (Evans *et al.*, 1983; Minshull *et al.*, 1989a, 1989b; Murray and Kirshner, 1989). The decrease in cyclin concentrations is due to the periodic degradation of cyclins. The cyclins contain a conserved sequence of amino acids known as the destruction box (Nugent *et al.*, 1991). The destruction box is the target of the ubiquitin-mediated proteolytic pathway that is responsible for the rapid degradation characteristic of this family (Glotzer *et al.*, 1991; King *et al.*, 1996). The ability to mutate the destruction box such that degradation is prevented has created a powerful tool, nondegradable cyclin, that has been used to discover many of the molecular behaviors involved in the cell cycle.

In contrast with the cyclins, protein levels of cdc2 are essentially constant throughout the cell cycle (Simanis and Nurse, 1986; Labbe *et al.*, 1989). Although cdc2 levels are constant, the kinase activity of this protein is periodic due to its regulation by cyclin and a series of phosphorylation events. Cyclins bind cdc2 to form the inactive MPF heterodimer

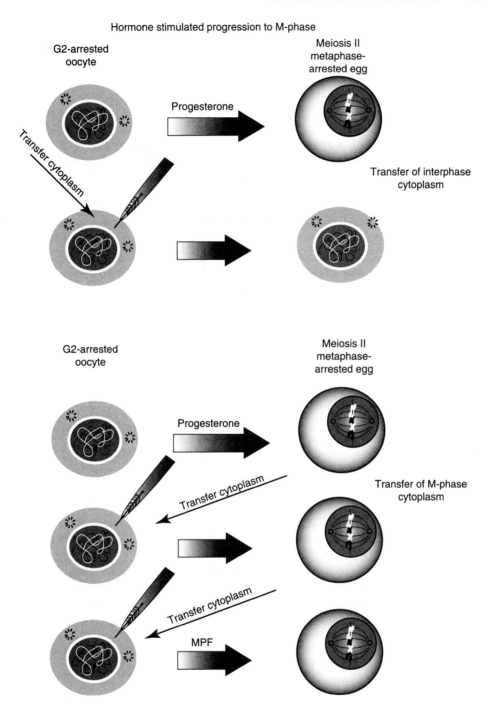

Figure 6.1 *Schematic of microinjection studies in oocytes that led to the characterization of MPF. Cytoplasmic material taken from oocytes in the first mitotic cycle when injected into meiotically arrested oocytes led to the progression of these cells to M-phase.*

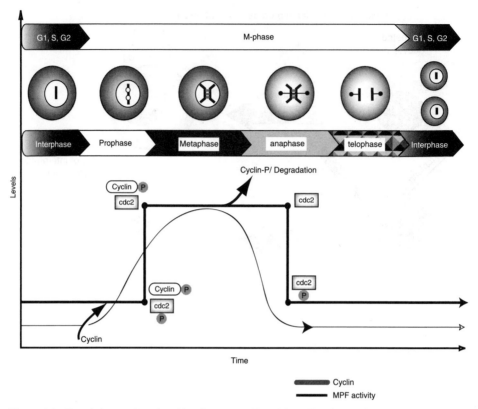

Figure 6.2 *Protein interactions that drive the progression of the cell cycle are shown in the upper half of the figure. Each reaction leads to the production of a new variable state. The corresponding level of MPF activity is shown below the molecular interactions. Solid arrows indicate a biological reaction, whereas dashed lines indicate the molecule participates as a cofactor.*

complex. Activation of cdc2 kinase (MPF) *in vivo* occurs once cyclin reaches a concentration threshold. A temporal lag of ∼10–20 minutes between cyclin reaching its threshold concentration and the full activation of MPF was found in *Xenopus* extracts by Solomon *et al.* (1990).

One purpose of the model we are going to examine is to determine what parameters are important to the time lag that is seen between cyclin reaching its threshold concentration and detectable activation of MPF. Solomon *et al.* (1990) explored the possibility of the lag being due to the rate of binding between cyclin and cdc2. However, he found the kinetics, 5 minutes for complete binding, to occur more rapidly than the lag and therefore an insufficient explanation for the delay (Solomon *et al.*, 1990). We will see later as we examine a modified version of a model put forth by Goldbeter (1991) that temporal delays can arise from the enzymatic properties and protein concentrations of the system.

Cyclins and cdc2 *in vivo* are found in multiple states. These states correspond with changes in kinase activity and to different stages of the cell cycle. Newly synthesized cyclin is unbound monomer cyclin. During the onset of interphase, unbound cyclin is the most abundant cyclin state. Cyclin binds cdc2 to form a cyclin-cdc2 complex, the inactive MPF. Inactive MPF accumulates throughout interphase as the total concentration of cyclin increases (i.e., bound + unbound = total). Although the total amount of cyclin increases throughout interphase and decreases during exit from mitosis, cyclin is distributed between the bound and unbound states. Nonetheless, the activation of MPF is often characterized in relation to changes in total concentration of cyclin.

The phosphorylation state of cdc2 is critical to the activation or inactivation of MPF. When bound to cyclin, cdc2 is phosphorylated and dephosphorylated to activate the MPF kinase. Together, phosphorylation and dephosphorylation of cdc2 kinase occurs within 20 minutes in *Xenopus* extracts generated by Solomon *et al.* (1990) and within 50 minutes in sea urchin eggs (Meijer *et al.*, 1991). We know from experimentation that cdc2 is phosphorylated on both a threonine and tyrosine. The latter is an inhibitory phosphorylation. When cdc2 is phosphorylated on tyrosine and bound to cyclin, MPF is inactive regardless of the phosphorylation state of threonine. MPF mediates its own activation by stimulating dephosphorylation of tyrosine in the inactive MPF (Gautier *et al.*, 1991; Kumagi and Dunphy, 1992). This creates an autocatalytic activation loop. Together, cyclin and cdc2 kinase create multiple variable states with two categories of functionality: active or inactive (Fig. 6.2).

Active cdc2 kinase also stimulates the degradation of cyclin (Felix *et al.*, 1990). MPF stimulates the ubiquitination of cyclin, and the ubiquitin proteolytic pathway is responsible for the rapid degradation that is required for the inactivation of MPF. This provides a negative feedback loop for the biological system. The kinetics are characterized by a temporal delay between high levels of cdc2 kinase activity and the degradation of cyclin. Felix *et al.* (1990) and Glotzer *et al.* (1991) observed the delay to be ~15 minutes. The delayed onset of cyclin degradation suggested that an intermediate factor must be involved to trigger the degradation of cyclin. The literature often referred to the intermediate factor as factor X (Goldbeter, 1991). We now know that ubiquitin activating proteins and the anaphase promoting complex (APC) mediate MPF-triggered cyclin degradation (King *et al.*, 1996). The degradation of cyclin leads to the inactivation of MPF. This completes the feedback loop between cyclin activation of MPF and MPF degradation of cyclin and subsequent inactivation of MPF.

6.1.2 Behavior and Experiment Highlights

The cell cycle behavior in *Xenopus* embryos is repetitious oscillations. The network of cell cycle reactions contains two commonly known structures for biochemical oscillations: autocatalytic activation and a negative feedback loop. Although the network of reactions is a prerequisite for oscillations, they are not sufficient. The dynamics are dependent on the parameters associated with each variable and the rate equations: V_{max}, initial concentrations, and K_m's.

Highlight 6.1

Expected Behaviors

Cyclin accumulates throughout the cell cycle.
MPF is activated after cyclin reaches a threshold concentration level.
There is a time delay/lag between MPF activation and cyclin degradation.
Delay exists between cyclin reaching threshold and activation of MPF.
Delay exists between activation of MPF and degradation of cyclin.

From the introduction, we can highlight some overall general behaviors. Cyclin accumulates linearly until its degradation is triggered by MPF. There is a temporal delay of $\sim$20 minutes between cyclin accumulation and the activation of MPF. Tyrosine phosphorylation and dephosphorylation occurs within 20 minutes of cdc2 incubation with saturating levels of cyclin (Solomon *et al.*, 1990; Meijer *et al.*, 1991). Although the rate of degradation is rapid, the onset of cyclin degradation does not occur immediately with the addition of active cdc2 kinase but rather has a delay of $\sim$15 minutes (Felix *et al.*, 1990; Glotzer *et al.*, 1991; King *et al.*, 1996). The model we explore in this chapter enables us to examine the relationship of thresholds to the length of the delay.

6.2 MODELING THE CELL CYCLE

Based on the experimental patterns identified above, a number of conceptual models have been developed and put forth. We can discuss conceptually the dynamics of the cell cycle as changes in MPF activation, changes in cdc2 kinase activation and in cyclin expression, or changes in molecular binding and phosphorylation states (Fig. 6.3). These conceptual models look at different degrees of detail. The appropriateness of a model will depend on what aspect of the biological phenomenon you are interested in. If you are interested in the effect of cdc2 phosphorylation on the kinetics of cell cycle progression, the first model that only grossly depicts MPF activation and inactivation would be insufficient. Conversely, if your focus requires knowing solely that

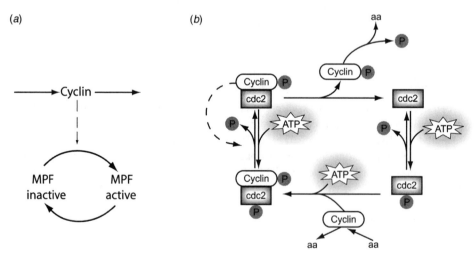

Figure 6.3 *Diagrams of MPF activation and inactivation at two different levels of abstraction. Each diagram illustrates the topological representation of distinct models of the cell cycle. Each model includes the activation and inactivation of MPF. They differ in their degrees of detail. (a). MPF is modeled as two states: MPF inactive and MPF active. Cyclin is produced and degraded, indicated by the solid arrows. The dashed line indicates cyclin's affect on the rate of MPF activation. (b). MPF is modeled at the level of protein interactions and phosphorylation states. This more detailed model allows rates of complex formation between cyclin and cdc2 to be accounted for explicitly.*

MPF is activated, the detailed dynamics of the binding rates of cdc2 and cyclin may be more cumbersome than helpful.

Some of the conceptual models have been transformed into mathematical models. The models were developed to explore the plausibility of mechanisms responsible for delayed cyclin degradation, oscillations, and changes in cell cycle duration (Goldbeter, 1991; Norel and Agur, 1991) or to explore whether the known set of biological factors (kinases, phosphatases) were sufficient to describe the full range of cell cycle behaviors (Tyson, 1991; Novak and Tyson, 1993). These models reproduce some or all of the behaviors observed in laboratory experiments. Each model reflects the core components of the cell cycle in various degrees of detail. At the level that we will be developing models, the basic components of the cell cycle engine are the same. The model will be used to explore two questions: Are feedback loops and thresholds sufficient to generate the periodic rise and fall in protein concentrations that are characteristic of the cell cycle patterns we see *in vivo*? Are the feedback loops and thresholds sufficient to generate delays in activation of MPF and degradation of cyclin?

6.2.1 System Statements of the Minimal

Let's begin with a simple model for the cell cycle engine. This model will embody the following statements:

1. Cyclin is synthesized and degraded.
2. MPF activation is dependent on cyclin concentrations.
3. MPF is activated and inactivated.
4. MPF activates cyclin degradation via a proteolytic enzyme.

This model can be used to explore whether feedback loops are sufficient for the generation of cycles and the observed temporal delay in activation of MPF and degradation of cyclin. The components are cyclin, MPF, and a proteolytic enzyme. MPF and the proteolytic enzyme in our model have two states, active and inactive. In this case, we describe the model as having three components. Thinking of molecules as a single factor with multiple states will become useful later for decreasing the number of variables to be calculated during our simulations. We can draw a diagram for the statements above as in Figure 6.4. This resembles earlier diagrams of MPF activity when it was not yet known that MPF is a complex of cyclin and cdc2. MPF in this model is synonymous with cdc2 kinase. However, the model will not explicitly model the molecular reactions of cyclin binding or phosphorylation and dephosphorylation. To avoid implying a false level of molecular detail, we will use the term *MPF*.

This model includes a negative feedback loop on the activation of MPF via the protease-mediated decrease in cyclin concentrations. We have chosen in this model to describe the relationship between cyclin and MPF phenomenologically, meaning that we ascribed a rate of MPF activation based on what is observed experimentally and that is proportional to total concentration of cyclin but that does not include the details of complex formation between cdc2 and cyclin.

As in previous chapters, we will give every solid arrow in our flowchart a symbolic notation that will be used in the differential equations that describe the model. There are six flows of material in the model that has been drawn. V_1 is the production of cyclin, V_2 is cyclin

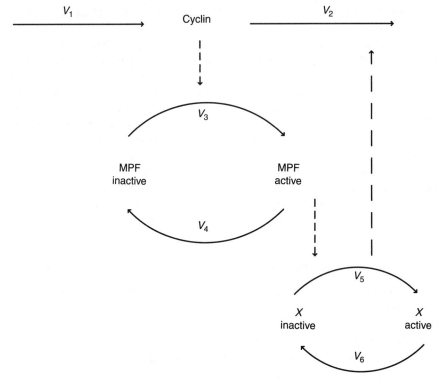

Figure 6.4 *The flow map indicates cyclin synthesis (V_1) and degradation (V_2). Cyclin acts as a cofactor (- - -) in the activation of cdc2 kinase (V_3) from an inactive to active state. Active cdc2 kinase becomes inactive (V_4). Active cdc2 kinase acts as a cofactor in the activation of the protease (V_5). The protease becomes inactive. The protease mediates cyclin degradation.*

degradation, V_3 is MPF activation, V_4 is MPF inactivation, V_5 is protease activation, and V_6 is protease inactivation. This diagram is similar to the schematic developed by Goldbeter (1991). The cyclin protease downstream of MPF is denoted by X. The dashed lines in the diagram indicate variables that act as modifiers or cofactors in the process.

6.2.2 Ordinary Differential Equations

We can now write a set of differential equations that describe the rate of change of each variable (cyclin, C; MPF active, M_a; MPF inactive, M_i; protease inactive, X_i; protease active, X_a) in terms of its input and output reactions.

1. Rate of change of cyclin = rate of cyclin production − rate of cyclin degradation

$$\frac{dC}{dt} = V_1 - V_2.$$

2. Rate of change of M_a = rate of M_i activation − rate of M_a inactivation

$$\frac{dM_a}{dt} = V_3 - V_4.$$

3. Rate of change of MPF$_i$ = rate of M_a inactivation − rate of M_i activation

$$\frac{dM_i}{dt} = V_4 - V_3.$$

4. Rate of change of X_a = rate of X_i activation − rate of X_a inactivation

$$\frac{dX_a}{dt} = V_5 - V_6.$$

5. Rate of change of X_i = rate of X_a inactivation − rate of X_i activation

$$\frac{dX_i}{dt} = V_6 - V_5.$$

This set of differential equations describes how each variable of the model changes over time. By relating to MPF as one factor with two different states (active and inactive) that is neither synthesized nor degraded, we can define the total concentration of MPF as constant and the inactive and active states as fractions of that constant value. We can reduce the number of differential equations by taking advantage of what we know about MPF and the protease X in terms of mass. Each factor exists in two states, active (M_a, X_a) and inactive (M_i, X_i) in the model. We can look at our diagram and see that there is no gain or loss of mass between M_a and M_i or X_a and X_i. If the amount of M_i increases, the amount of M_a must decrease by the same amount and vice versa. The same is true for X_a and X_i. Another indicator of mass conservation is that M_a and M_i are both defined by V_3 and V_4. X_a and X_i are both defined by V_5 and V_6. Because mass is conserved, we can write the amount of one variable in terms of the other and thereby reduce the set of equations from five to three (Table 6.1; see "Simplifying the set of ODEs" in Chapter 4).

6.2.3 Michaelis-Menten Rate Equations

The exact behavior of this model depends on the rate laws used to define the velocities in the differential equations and the values assigned to the accompanying parameters. As modelers, we choose rate laws that we deem appropriate for the reaction we are modeling. In this case, we will work with Michaelis-Menten rate laws.

TABLE 6.1   Table of ODEs

1. $\dfrac{dC}{dt} = V_1 - V_2$

2. $\dfrac{dM_a}{dt} = V_3 - V_4$

3. $\dfrac{dM_i}{dt} = V_4 - V_3$

4. $\dfrac{dX_a}{dt} = V_5 - V_6$

5. $\dfrac{dX_i}{dt} = V_6 - V_5$

TABLE 6.2^a Rate Equations and Parameters

	Reactions	Parameters	
Cyclin	conc	0.01	
M_a	conc	0.01	
X_a	conc	0.01	
Cyclin synthesis			
k1	V1	k1	0.025
Cyclin degradation			
vmax2 * X_a * Cyclin/(Km2 + cyclin)	V2	vmax2	0.25
		Km2	0.02
MPF activation			
vmax3 * Cyclin/(Kc3 + Cyclin) * M_i/(Km3 + M_i)	V3	vmax3	3
		Km3	0.005
		Kc3	0.5
MPF inactivation			
vmax4 * M_a/(Km4 + M_a)	V4	vmax4	1.5
		Km4	0.005
Protease activation			
vmax5 * M_a * X_i/(Km5 + X_i)	V5	vmax5	1
		Km5	0.005
Protease inactivation			
vmax6 * X_a/(Km6 + X_a)	V6	vmax6	0.5
		Km6	0.005

^{a}Parameter values in the table match those used in the Stella concept map. Vmax is used to indicate Vmax values for reactions. Km is the Michaelis-Menten constant Km for the reaction. However, Kc is used for the Km governing cyclins behavior in MPF activation V# is the overall reaction rate.

Michaelis-Menten takes into account that the enzyme may become saturated. Once saturated, the rate of the reaction no longer increases linearly with increasing substrate concentrations. Michaelis-Menten rate laws determine the velocity in terms of the concentration of the substrate, the V_{max} for the reaction, and Michaelis constant K_m. From the set of rate laws, we can see the parameters for which we need to provide values (Table 6.2). The Michaelis-Menten rate laws describe the behavior of enzymatic reactions better than mass action kinetics, particularly when the reaction has saturation characteristics. The values shown in Table 6.2 are those used by Goldbeter (1991). These values produce oscillations in cyclin, MPF, and protease activity that closely mimic the cell cycle patterns of *Xenopus* embryos.

6.2.4   Parameter Values: Lab and Model

An important discussion is how to think about the units or lack of units that exist in the models we are presenting and the data that is collected in the laboratory. Where do the concentration, V_{max}, and K_m values come from or how are they estimated? One of the initial relationships to discuss is that of concentrations. In the simulations performed in Stella, the initial concentrations were set to 0.01. Why? One reason is that it is twofold greater than the K_m values that were used in the model. K_m was set to 0.005. By the initial concentrations being greater than the K_m, the reactions will run closer to the enzymatic maximum velocity. Goldbeter has previously shown that sharp thresholds appear in the dynamics of coupled converter enzymes (i.e. phosphatase and kinase) when they are saturated (Goldbeter, 1991). By the concentrations being greater than the K_m, the model system is optimized for discovering if threshold conditions exist and how they affect the behavior of the system.

The concentrations of active MPF (M_a) and active protease (X_a) are unitless because they have been"scaled" by the total concentration of MPF and protease, respectively. The total concentrations of MPF and protease within the model are conserved (see Chapter 4). No mass is removed from the system for these two variables. Instead, the proteins transit between active and inactive states, such that the amount of protein in either state is a fraction of the total concentration, which is scaled to 1. The experimentally derived concentration for the total amount of each variable can be multiplied by the parameter value in the simulation in order to re-establish scales. Goldbeter (1991) assumed a concentration of cdc2 kinase (MPF in our model) to be 4 μM based on Sea Urchin studies (Labbe *et al.*, 1989). The distribution of MPF between active and inactive states is thus the fraction seen in the model multiplied by the parameter 4 μM.

From laboratory experiments we have ranges of concentration values that have been found for cyclin and cdc2 kinase *in vivo* and *in vitro*. Solomon *et al.* (1990) worked primarily with 70 nM cyclin in *Xenopus* extracts although 32 nm was sufficient to trigger activation of H1 kinase activity; Sha *et al.* (2003) explored the concentrations of cyclin that allow for exit from mitosis in *Xenopus* extracts and found a threshold of 40 nm. It is believed that cyclin concentrations are much smaller than the total concentration of cdc2 kinase (Labbe *et al.*, 1988, 1989; Meijer *et al.*, 1989; Solomon *et al.*, 1990).

We can estimate a rate constant for reactions based on the experimental data by finding the slopes of lines from linear or exponential functions. If we assume exponential decay or growth, we need to know either the observed half-life of the substrate molecule or the amount that is lost from one time point to the next. When we know the half-life, we set the observed rate equal to the natural log of 2 over the rate constant (k) and solve for k. This rate constant is then treated as the V_{max} of the Michaelis-Menten equation.

Degradation rates have been calculated and observed by Glotzer *et al.* (1991) to have an 0.46/min ubiquitin conjugation rate and a shorter rate of 0.053/min for degradation of ubiquitin conjugates. This observed rate is almost twofold faster than what was used as a variable by Goldbeter (1991) for the maximum velocity for cyclin degradation. We can implement the model and then test the impact of the different rates on the overall behavior of the system.

6.3 IMPLEMENTATION IN STELLA

The described model has been implemented in Stella with Michaelis-Menten rate equations (Fig. 6.5). The parameters of the rate laws are drawn as converters (O) to make them readily visible to the reader. The simulation results are shown in Figure 6.6. The smallest parameter value is 0.005. This same value was used as the step size for the simulation. By running the simulation at both 0.005 and 0.0025 with the same method, Euler, we can confirm that this is an appropriate step size for approximating the differential equations that describe the system's behavior. Identical results were obtained using the same step size with an alternative approximation method (Runge-Kutte 2; see Chapter 4).

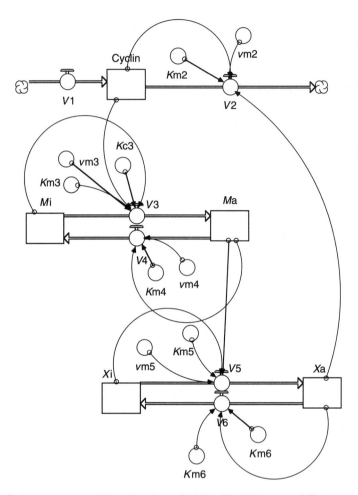

Figure 6.5 *Stella concept map of the cell cycle model in the "World" view panel. Stocks (▢) represent the pool of proteins, flows (⚬—⚬—⚬) represent reactions or processes, converters (◯) represent parameters, such as K_m and V_{max}, of the reaction rates, and connectors indicate that a stock or converter participates in the reaction.*

6.4 SIMULATION RESULTS

It is worth noting that as in any experiment, we must be careful that what is inferred from the results can be deduced from the experimental system. The variables we have—cyclin, M_i, M_a, X_a, and X_i—are theoretical markers for the concentrations of proteins and in our case the concentration of proteins with particular capabilities. An increase in M_a, active MPF, is an increase in the population of MPF molecules that are active. It is not a direct measure of activity. The only measure of MPF activity in the model is in the rate at which active protease (X_a) is formed. In this case, MPF activity is described by a V_{max} and Michaelis constant that define how the amount of protease changes in relation to MPF. This is a different measure than what we see in laboratory experiments where MPF activity is defined either in relation to H_1 kinase assays or the ability to

1: Cyclin   2: Ma   3: Xa

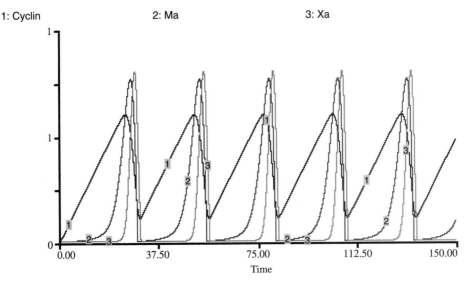

Figure 6.6 *Simulation results of cell cycle model based on parameter and concentration values provided in Table 6.1. Simulation was run with Euler approximation method with delta t (DT) set to 0.005. Graph was generated in Stella 7.0 Research.*

stimulate 50% of a population of cells to progression to meiosis or mitosis (Wu and Gerhart, 1980).

We can see from the simulation results that characteristic cycle behaviors are obtained (Fig. 6.6). We see that the periodicity of the simulation results is ∼30 minutes. Embryonic cell cycles *in vivo* are approximately 30 minutes and *in vitro* extracts vary between 35 and 55 minutes (Hara *et al.*, 1980; Murray and Kirshner, 1989). Similar to the protein patterns associated with interphase and the onset of mitosis, cyclin concentrations increase prior to increases in active MPF. The onset of anaphase correlates to increasing amounts of active protease and rapid decline in cyclin concentrations. In the model, cyclin concentrations only decrease after activation of the protease.

In addition to cyclin concentrations increasing prior to activation of MPF, we see that over the first 30 minutes of simulated time, cyclin levels rise linearly with a constant slope. In contrast, levels of active MPF are low in the first 14 minutes then rise very rapidly with an increasing slope until it peaks at 28 minutes. There is not a linear relationship between MPF activation and cyclin concentrations. Also, we do not see in this graph (Fig. 6.6) the pattern seen by Solomon *et al.* (1990) that indicated the threshold behavior between cyclin and active MPF. This may be partly due to our ability to sample the data every minute in the simulation, achieving a higher resolution, whereas samples are likely taken at longer time intervals, e.g. every 10 minutes, in laboratory settings. We examine the threshold behavior further with our next simulation.

The amount of active protease (X_a) increases only after MPF is activated as expected. Given the activation of X occurs after MPF, there is a built-in time delay for the increased degradation of cyclin. Although the simulation began with cyclin at a concentration of 0.01, after the first cycle, cyclin levels decrease to 0.11 rather than 0.01. The lowest level of cyclin is determined by the concentration that brings the rate of degradation equal to the rate of synthesis, $V_1 = V_2$. The inactivation of MPF begins in the first cycle at 27 minutes and is complete by 30 minutes, such that the inactivation of MPF is rapid, occurring within 3 minutes of the peak activation level.

6.5   NONDEGRADABLE CYCLIN AND CYCLOHEXAMIDE TREATMENT

Let's now vary this model to test its ability to reproduce experimental results. By making the cyclin concentration constant throughout the simulation, we can examine the behaviors of the protease and MPF as a function of particular cyclin concentrations. To achieve this, we have removed the influx and degradation of cyclin (Fig. 6.7). This mimics the conditions created by adding nondegradable cyclin in the presence of cyclohexamide to inhibit synthesis of new proteins. It is under these conditions that threshold behavior is highlighted. In order to determine the threshold concentration in the simulated system, we chose cyclin concentration values that temporally correlated with the steep rise in MPF activity. When we run simulations with fixed values of cyclin, we see levels of active MPF plateau at some maximum value (Fig. 6.8). Lower concentrations of cyclin, 0.26 through 0.45, resulted in a small amount of MPF (<0.09) becoming active. Full activation of MPF (98%) was obtained with cyclin concentration of 0.6. When we plot the maximum amounts of active cdc2 kinase against cyclin concentration with which it was obtained, we see the same steep threshold curve as seen by Solomon *et al.* (1990) (Fig. 6.8).

6.6   DISCUSSION OF MODELS

In the minimal mitotic model implemented in this chapter, the delay between MPF activation and cyclin threshold is 3–4 minutes. The delay is determined in part by the rate of activation of the protease and by our definition of "delay." Here we define delay as the time between a cyclin threshold of 0.47 and half the maximum level reached by

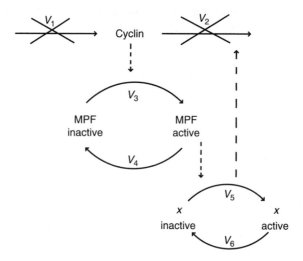

Figure 6.7   *Interaction map of model, modified to represent reactions no longer included in simulation. This was implemented within Stella by setting reaction rates to zero for the indicated processes.*

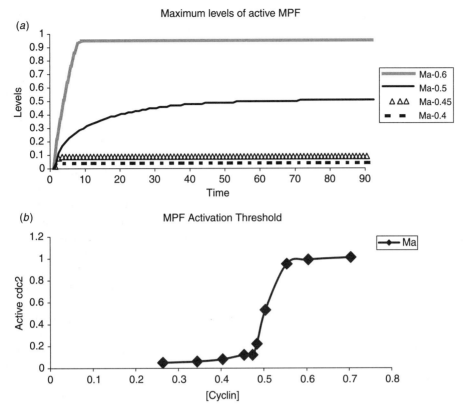

Figure 6.8 *(a). Levels of active MPF (Ma) over time are plotted from four simulations. Cyclin initial condition values were fixed in each simulation at 0.4, 0.45, 0.5 or 0.6. No oscillatory behavior is seen in these simulations. (b). Maximum levels of active MPF are plotted relative to the fixed cyclin levels in the simulation. Cyclin levels were fixed over a range of 0.24 to 0.7. The graph was generated in Excel by importing numerical data points from Stella.*

active MPF. This threshold value is not what is expected *in vivo* but is rather a product of the parameter values chosen for the simulation.

This model does not take into account MPF's positive feedback loop to its own activation. Neither does it explicitly model regulation of MPF by the phosphatase, Cdc25 and the inhibitory kinase, Wee1. Tyson (1991) and Novak and Tyson (1993) took both of these aspects into account in their models of a general cell cycle machine and the *Xenopus* embryo, respectively. Novak and Tyson (1993) have shown that the delayed activation of cdc2 kinase is dependent on the relative activity of Wee1 and Cdc25. The level of active cdc2 kinase, which changes the amount of active Wee1 and Cdc25 and therefore the effective rates of MPF inactivation and activation, respectively. By incorporating Wee1 and Cdc25 into the model, we expect the delay between threshold and activation to more accurately reflect the kinetics of what is seen experimentally.

A single protease with the parameter values used in this model is insufficient to generate the same length of delay in the onset of degradation (Felix *et al.*, 1990; Glotzer *et al.*, 1991; Marlovits, 1998). Subsequent models have included upstream activators of the ubiquitin pathway, APC, and show delayed onset of degradation of 15–20 minutes (Novak and Tyson, 1993). In our model, we include the activation of a protease that degrades cyclin. The degradation rate is set to be linearly proportional to the amount of active X. If there is no active protease, no degradation occurs. Our initial concentration of X_a is 0.01, a sufficient quantity for a low level of degradation to occur (data not shown). Once pronounced degradation begins, the rate is proportional to the amount of available X_a.

The model simplifies many currently known biological events. Binding of cyclin to MPF is not described. Instead, we model the effect of increased cyclin concentrations on MPF activation. In order to take cyclin binding to MPF into account, we would modify the topology of the model in both the schematic, for reference, and in Stella, for implementation. Tyson (1991) and Novak and Tyson (1993) have generated models that explicitly include the binding reactions between molecules. In their models, the rate equations are mass action kinetics rather than Michaelis-Menten. They did not include the feedback loops as seen in our model here but rather model MPF activating MPF. We do not address these models here because of the larger number of variables and parameters. By increasing the number of details, the number of variables and parameters that are investigated also increase. Although additional details may provide a more "realistic" context, they also increase the number of dimensions that can be varied or need to be analyzed in relation to the potential results. Additional skills in manipulating differential equations and phase plane analysis are desirable to effectively analyze and interpret the behavior of such models.

The original models of Tyson (1991) and Novak and Tyson (1993) have been expanded to examine regulators of cell cycle core components by Wee1 and Cdc25 and also connected these underlying mechanisms to definitions of division, size, and DNA replication. These models have been used to predict the size and kinetics related to mutant Wee1 proteins in yeast. Recently, Sha *et al.* (2003) experimentally confirmed threshold predictions by Novak and Tyson (1993) for the inactivation of MPF and exit from mitosis. Given many parameters are unknown for the model components, these authors took advantage of phase plane analysis to discover the behavior of the system and parameter values that reproduce the results seen in laboratory observations. This method of analysis is a powerful means of examining the qualitative behavior of variables in relation to one another as a function of their differential equations.

6.7   CONCLUSION

The model in this chapter is a simplified version of cell cycle molecular interactions. These simplifications take into account the major properties of cyclin synthesis, cyclin activation of cdc2 kinase, activation of a protease, and the degradation of cyclin. Each aspect of the model could be expanded to take into account additional details: cyclin binding to cdc2 kinase, phosphorylation and dephosphorylation steps, and the ubiquitin degradation pathway. The purpose of this chapter was to introduce the reader to modeling the basic assumptions of how the cell cycle works. With this basic understanding, it should now be possible to adapt this model or others to explore questions in one's own systems.

The model here shows that the presence of delayed negative feedback combined with thresholds and time delays that occur as a function of the relationship between both cyclin:cdc2 kinase and cdc2 kinase:protease are sufficient for the generation of cell cycle oscillations (Goldbeter, 1991). The role of cdc2 autocatalysis has been explored in other models and is also known to be sufficient for the generation of cell cycle oscillations. These models do not have to be mutually exclusive but may rather be a method of ensuring the ability for the cell cycle to proceed. An important feature of the cell cycle that is not explored in the model is the regulatory molecules that exert negative and positive controls on cell cycle progression. These factors have been explored in more detailed models (Novak and Tyson, 1993; Marlovitis *et al.*, 1998; Chen *et al.*, 2000). Given these models introduce additional variables and parameters, many values are absent, and the modelers resort to phase plane analysis to discover parameter values that create the observed experimental behaviors. The parameter values that solve the equations become a premise for interrogating the experimental data. Are substrate concentrations found *in vivo* that are seen in the model? Are the rate constants used biologically significant or consistent with experimental results?

A key issue is the absence of V_{max} and K_m values for the enzymes involved. Modelers will often use a method of parameter estimation to find parameter values that create the behaviors observed in the experimental system. Although this allows us to say that the model can produce the behavior, it is less satisfying than using experimentally derived parameter values. In the model developed here, values were borrowed from multiple experimental systems and estimated values for K_m and V_{max}. On one hand, the ability to take data from multiple experimental systems and obtain reasonable results in the model suggests that the model is universal. On the other hand, data from one experimental system can be a better test of whether the model sufficiently describes the biological system and leaves less room for questioning sources of variation.

BIBLIOGRAPHY

Chen KC, Csikasz-Nagy A, Gyorffy B, *et al.* (2000). Kinetic analysis of a molecular model of the budding yeast cell cycle. *Molecular Biology of the Cell* 11(1):369–391.

Chen KC, Calzone L, Csikasz-Nagy A, *et al.* (2004). Integrative analysis of cell cycle control in budding yeast. *Molecular Biology of the Cell* 15(8):3841–3862.

Doree M (1990). Control of M-phase by maturation-promoting factor. *Current Opinion in Cell Biology* 2(2):269–273.

Evans T, Rosenthal ET, Youngblom J, *et al.* (1983). Cyclin: a protein specified by maternal mRNA in sea urchin eggs that is destroyed at each cleavage division. *Cell* 33(2):389–396.

Felix MA, Labbe JC, Doree M, *et al.* (1990). Triggering of cyclin degradation in interphase extracts of amphibian eggs by cdc2 kinase. *Nature* 346(6282):379–382.

Gautier J, Minshull J, Lohka M, *et al.* (1990). Cyclin is a component of maturation-promoting factor from *Xenopus*. *Cell* 60(3):487–494.

Gautier J, Solomon MJ, Booher RN, *et al.* (1991). Cdc25 is a specific tyrosince phosphatase that directly activates p34cdc2. *Cell* 67(1):197–211.

Glotzer M, Murray AW, Kirschner MW (1991). Cyclin is degraded by the ubiquitin pathway. *Nature* 349(6305):132–138.

Goldbeter A (1991). A minimal cascade model for the mitotic oscillator involving cyclin and cdc2 kinase. *Proceedings of the National Academy of Sciences USA* 88(20):9107–9111.

Hara K, Tydeman P, Kirschner M (1980). A cytoplasmic clock with the same period as the division cycle in *Xenopus* eggs. *Proceedings of the National Academy of Sciences USA* 77(1):462–466.

Hyver C, Le Guyader H (1990). MPF and cyclin: modelling of the cell cycle minimum oscillator. *Biosystems* 24(2):85–90.

King RW, Glotzer M, Kirschner MW (1996). Mutagenic analysis of the destruction signal of mitotic cyclins and structual characterization of ubiquitinated intermediates. *Molecular Biology of the Cell* 7(9):1343–1357.

Kumagai A, Dunphy WG (1992). Regulation of the cdc25 protein during the cell cycle in Xenopus extracts. *Cell* 70(1):139–151.

Labbe JC, Picard A, Karsenti E, *et al.* (1988). An M-phase-specific protein kinase of Xenopus oocytes: partial purification and possible mechanism of its periodic activation. *Developmental Biology* 127:157–169.

Labbe JC, Picard A, Peaucellier G, *et al.* (1989). Purification of MPF from starfish: identification as the H1 histone kinase p34cdc2 and a possible mechanism for its periodic activation. *Cell* 57(2):253–263.

Lee TH, Solomon MJ, Mumby MC, *et al.* (1991). INH, a negative regulator of MPF, is a form of protein phosphatase 2A. *Cell* 64(2):415–423.

Lorca T, Castro A, Martinez AM, *et al.* (1998). Fizzy is required for activation of the APC/cyclosome in *Xenopus* egg extracts. *EMBO Journal* 17(13):3565–3575.

Marlovits G, Tyson CJ, Novak B, *et al.* (1998). Modeling M-phase control in *Xenopus* oocyte extracts: the surveillance mechanism for unreplicated DNA. *Biophysical Chemistry* 72(1–2):169–184.

Masui Y, Markert C (1971). Cytoplasmic control of nuclear behavior during meiotic maturation of frog oocytes. *Journal of Experimental Zoology* 177(2):129–145.

Meijer L, Azzi L, Wang JY (1991). Cyclin B targets p34cdc2 for tyrosine phosphorylation. *EMBO Journal* 10(6):1545–1554.

Meijer L, Arion D, Golsteyn R, *et al.* (1989). Cyclin is a component of the sea urchin egg M-phase specific histone H1 kinase. *EMBO Journal* 8(8):2275–2282.

Minshull J, Blow JJ, Hunt T (1989a). Translation of cyclin mRNA is necessary for extracts of activated xenopus eggs to enter mitosis. *Cell* 56(6):947–956.

Minshull J, Pines K, Golsteyn R, *et al.* (1989b). The role of cyclin synthesis, modification and destruction in the control of cell division. *Journal of Cell Science* 12(suppl):77–97.

Morla AO, Draetta G, Beach D, *et al.* (1989). Reversible tyrosine phosphorylation of cdc2: dephosphorylation accompanies activation during entry into mitosis. *Cell* 58(1):193–203.

Murray AW, Hunt T (1993). *The Cell Cycle: An Introduction*. New York: Oxford University Press; 264p.

Murray AW, Kirschner MW (1989). Cyclin synthesis drives the early embryonic cell cycle. *Nature* 339(6222):275–280.

Norel R, Agur Z (1991). A model for the adjustment of the mitotic clock by cyclin and MPF levels. *Science* 251:1076–1078.

Novak B, Tyson JJ (1993). Numerical analysis of comprehensive model of M-phase control in *Xenopus* oocyte extracts and intact embryos. *Journal of Cell Science* 106:1153–1168.

Nugent JH, Alfa CE, Young T, *et al.* (1991). Conserved structural motifs in cyclins identified by sequence analysis. *Journal of Cell Science* 99:669–674.

Nurse P (1990). Universal control mechanism regulating onset of M-phase. *Nature* 344:503–508.

Sha W, Moore J, Chen K, *et al.* (2003). Hysteresis drives cell-cycle transitions in *Xenopus laevis* egg extracts. *Proceedings of the National Academy of Sciences USA* 100(3):975–980.

Simanis V, Nurse P (1986). The cell cycle control gene *cdc2+* of fission yeast encodes a protein kinase potentially regulated by phosphorylation. *Cell* 45:261–268.

Solomon MJ, Glotzer M, Lee TH, *et al.* (1990). Cyclin activation of p34cdc2. *Cell* 63(5):1013–1024.

Tyson JJ (1991). Modeling the cell division cycle: cdc2 and cyclin interactions. *Proceedings of the National Academy of Sciences USA* 88(16):7328–7332.

Wu M, Gerhart JG (1980). Partial purification and characterization of maturation promoting factor from eggs of Xenopus laevis. *Developmental Biology* 79:465–477.

Chapter 7

Calcium Dynamics

The goal of this chapter is to examine how spatiotemporal modeling can enhance our understanding of the cellular mechanisms that determine calcium dynamics. To do this, we will first review what is known about the mechanisms involved in calcium dynamics. We will then use the N1E-115 neuroblastoma cell line as our example of a biological system for which to develop a conceptual and computational model. Diagrams will be used to facilitate our creation of the conceptual model and initial corresponding mathematical descriptions. We then work with Virtual Cell to discuss and implement characteristics of our model, particularly spatial features. We are choosing to use Virtual Cell because it is one of a few tools that model spatial data and has a user interface tailored for biological models. As well, models of calcium dynamics have been studied, published, and made available with this tool.

7.1 THE BIOLOGY OF CALCIUM

Calcium is a well studied ion that participates in many aspects of cellular physiology such as: cellular signaling, nerve impulses, fertilization, and other cellular behaviors. A number of fluorescent microscopy techniques are designed to measure relative and quantitative amounts of the calcium in the cytosol and organellar compartments. By monitoring levels of calcium over time, we obtain the empirical data that describes its spatiotemporal behaviors—dynamics. Examples of calcium patterns include calcium oscillations, spikes, and waves (O'Sullivan *et al.*, 1989; Rooney *et al.*, 1990; Berridge, 1993; Kasai *et al.*, 1993). Localized and global changes in calcium levels lead to many subsequent events such as secretion, contraction, motility, and embryonic development.

Calcium concentration and distribution in the cytosol and organelles are regulated by the actions of calcium pumps, channels, and by binding proteins (Fig. 7.1). The Na-Ca$^+$

A Cell Biologist's Guide to Modeling and Bioinformatics. By Raquell M. Holmes
Copyright © 2007 John Wiley & Sons, Inc.

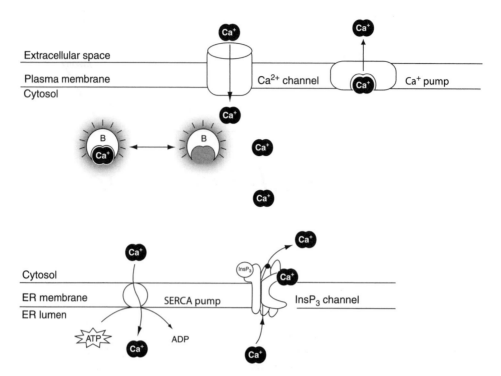

Figure 7.1 *Schematic of molecular components of cytosolic calcium dynamics including channels and pumps at plasma membrane and endoplasmic reticulum.*

exchanger at the plasma membrane, ATP pumps in membranes of the endoplasmic and sarcoplasmic reticula, and the uniporter within the mitochondria move calcium out of the cytosol to the extracellular environment, endoplasmic and sarcoplasmic reticula, or mitochondria, respectively (Gill *et al.*, 1981; Gill and Chueh, 1985). These mechanisms remove calcium from the cytosol. In contrast, calcium channels in both the plasma membrane and endoplasmic reticular (ER) membranes release calcium into the cytosol creating an influx of calcium. Together, the influx and efflux of calcium determine the amount of calcium within the cytosol and organelles.

Calcium in the cytosol can be classified as existing in two states: bound and unbound to buffers. Calcium binding proteins, as a population within the cell, buffer the amount of cytosolic calcium free to diffuse through the cytosol or free to interact with other cellular molecules. Unbound calcium is often referred to as *free calcium*; the population of binding proteins and molecules as calcium buffers. Free calcium is available to bind to fluorescent indicators (e.g., Fura-2, indo-1), used to measure calcium concentrations and detect calcium waves and oscillations.

At a very gross level, the features that determine calcium concentrations in a cell generally include influx and efflux of calcium through pumps and channels and buffering and storage in cellular compartments. The distribution of calcium within the cytosol and organelles is a factor that affects localized changes in concentrations of calcium. In the next few pages of this chapter, we discuss a generalized model of InsP$_3$-regulated calcium dynamics and then tailor the model to specifically examine calcium dynamics in the neuroblastoma cell including spatial distributions of proteins and organelles.

7.2 MECHANISM RESPONSIBLE FOR CALCIUM BEHAVIOR

Two mechanisms are responsible for maintaining low, $\sim 7 \times 10^{-5}$ μM, cytosolic concentrations of free calcium, despite high concentrations of calcium in the extracellular space, $\sim 1-2$ mM. They are (1) extrusion from the cell or sequestering into organelles and (2) buffering of calcium by calcium binding proteins. Increases in cytosolic calcium concentrations are attributed to influx of extracellular calcium and release of intracellular calcium stores. Spatial distributions of cellular machinery have been proposed as the basis for localized differences in calcium concentrations within the cell (Fink *et al.*, 1998).

7.2.1 Extrusion and Sequestering

ATP pumps, uniporters, and ion exchangers are the dominant mechanisms for removing calcium from the cytosol (Fig. 7.2). ATP pumps in the plasma membrane, sarcoplasmic and endoplasmic membranes undergo a cycle of phosphorylation and dephosphorylation that results in a single calcium ion being transported across the membrane. The activity

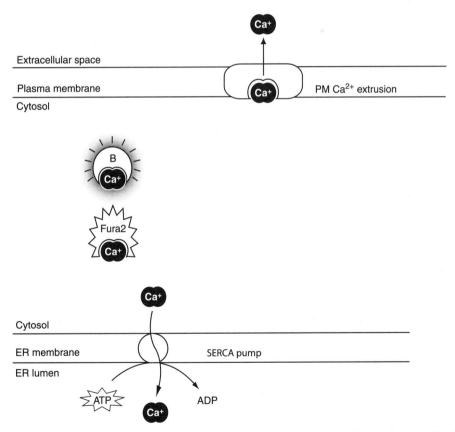

Figure 7.2 *Calcium extrusion and sequestration. Schematic of molecular mechanisms responsible for decreasing free cytosolic calcium concentrations. Plasma membrane pumps have been drawn as a single pump. Calcium binding with endogenous buffers (B) decreases the concentration of free calcium. Sarco-endoplasmic reticular ATPase pumps calcium into the endoplasmic reticula.*

of these pumps is best described by Hill-type kinetics (Gill *et al.*, 1981; Gill and Chueh, 1985; Lytton, 1992). Plasma membrane calcium pumps differ from intracellular pumps in location and sensitivities to pharmacologic agents. The intracellular pumps are known as the sarco-endoplasmic reticulum calcium ATPase, or SERCA, pumps. Both types of pumps help maintain low calcium concentrations in the cytosol by removing calcium from the cytosol. The pumps and Na^+/Ca^{2+} exchanger in the plasma membrane remove calcium from the cytosol by extruding it from the cell. In contrast, the SERCA pumps remove calcium by sequestering the calcium within reticular structures. In muscle cells, this would be the sarcoplasmic reticulum; in other cells, the endoplasmic reticulum.

7.2.2   Cytosolic Calcium Buffers

Buffers play a significant role in the characteristics of the spatiotemporal patterns of calcium. Calcium buffers are essentially all those proteins that interact with cytosolic calcium and thus remove it from the "free" pool (Fig. 7.2). Less calcium exists free in the cytosol when buffers bind calcium tightly. Conversely, when buffers bind calcium loosely, more calcium is free. The concentration of total cytosolic buffer has been calculated to be $100-410$ μM (Allbritton *et al.*, 1992; Milner *et al.*, 1992; Klingauf and Neher, 1997). Some measurements indicate that in the cytoplasm, 99 out of 100 calcium ions are bound by mobile and immobile factors (Neher and Augustine, 1992; Tse *et al.*, 1994).

Fluorescent calcium indicators are used to visualize calcium dynamics. These chelating agents act as additional buffer. The use of fluorescent calcium indicators can affect the dynamics we are studying by changing the speed, rate, or amplitude of the calcium patterns. This was seen in the work of Nuccitelli *et al.* (1993) with *Xenopus* oocytes where exogenous buffers altered the speed of calcium waves. The introduction of the fluorescent indicators is the introduction of an exogenous buffer to the biological system. Thus, fluorescent indicators become an important parameter to take into account when creating models of experimental studies.

7.2.3   Influx of Calcium to the Cytosol

Cytoplasmic calcium levels increase in response to receptor signaling, nerve pulses, or fertilization events due to calcium influx through channels at the plasma membrane or calcium release from organelles. The dominant mechanism of calcium influx to the cytoplasm is dependent on the nature of the stimulatory signal. In neuroblastoma cells, increases in calcium can be seen as early as 10 seconds. Yet, influx of calcium through plasma membrane channels occurs after $\sim$30 seconds (Iredale *et al.*, 1992). The early rise in calcium is believed to be due to the release of calcium from internal stores. In this chapter, we focus on release of intracellular calcium from the endoplasmic reticulum. The mechanisms for this release include $InsP_3$-stimulated calcium release from endoplasmic and sarcoplasmic reticula via $InsP_3$ receptor channels and calcium leakage (Fig. 7.3).

7.2.3.1   *InsP₃ Channel–Mediated Calcium Release* $InsP_3$, the second messenger generated downstream of G protein receptors via phospholipase C action on PIP2 (phosphatidyl-1,4 bisphosphate), regulates the release of calcium from reticular stores (Berridge and Irvine, 1984; Berridge, 1993). $InsP_3$ binds to a tetrameric $InsP_3$-receptor ($InsP_3R$) that is also a calcium channel in the membranes of endoplasmic

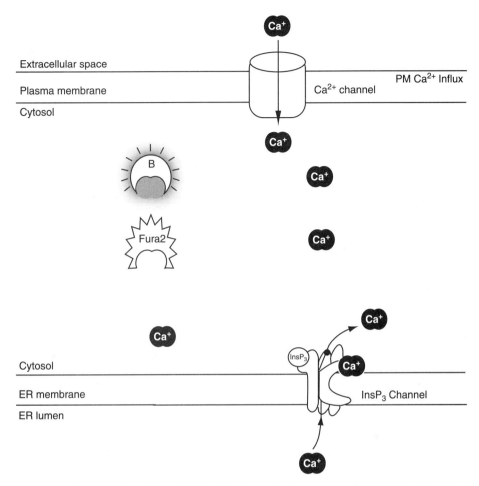

Figure 7.3 *Calcium influx. Schematic of molecular mechanisms responsible for influx of calcium to cytoplasm. Calcium travels down the concentration gradient through activated calcium channels in the plasma membrane. InsP₃ receptor channel in the endoplasmic reticula releases calcium when calcium and InsP₃ are appropriately bound.*

and sarcoplasmic reticula (Bezprovanny *et al.*, 1991). The InsP$_3$-receptor channel is both activated and inhibited by calcium. In order for the channel to open, calcium cannot be bound to the inhibitory site of the channel, while the activation sites must have bound InsP$_3$ and calcium. The probability of channels being open increases as calcium concentrations rise to 0.25 μM and decreases at higher concentrations when activating levels of InsP$_3$ are present and the calcium inhibition site is likely to be bound (Bezprozvanny *et al.*, 1991; De Young and Keizer, 1992). No ryanodine-sensitive receptors have been identified in the neuroblastoma cell (Wang *et al.*, 1995).

7.2.4 Spatial Distributions

A particularly important feature of calcium dynamics is the role of cellular structures and diffusion in the generation of cellular calcium patterns. In our previous chapters on cell cycle (Chapter 6) and metabolism (Chapter 5), we focused on the modeling of kinetic reactions in the absence of space. In other words, our differential equations and rates only took into account time and assumed that the concentrations of substrates and products were homogenous in space. However, calcium dynamics are characterized and influenced by complex and diverse spatial patterns.

Spatial characteristics of cells can influence the calcium dynamics. Diffusion rates of InsP$_3$ and calcium have been hypothesized to act as factors determining wave propagation (Jaffe, 1983, 1991; Meyer *et al.*, 1988; Parker and Ivorra, 1990; Lechleiter *et al.*, 1991; DeLisle and Welsh, 1992; Wang and Thompson, 1995). In some cases, it appears that calcium diffusion is the dominant determinant and in other cases the determinant is InsP$_3$. Diffusion is a spatial characteristic of the molecule within the 3D environment. We can characterize the spatial patterns of calcium in terms of where it initiates and what direction within a 3D space the changes travel. The 3D environment is the cellular architecture, which includes the overall geometry of the cell, distribution of organelles, and molecular species.

7.3   EXPERIMENTAL SYSTEM MODEL

The work of Fink *et al.* (2000) was chosen as our example system because they successfully use a computer model to generate experimentally testable hypotheses and have a breadth of support materials available on tutorial pages of the Web site for Virtual Cell (http://www.vcell.org). The tool, wealth of quantitative data, and resource materials provide a rich biological case study with which to work. The challenge for Fink and for us is organizing the preexisting knowledge of the biological system into wellstructured relationships and mathematical descriptions that can be used to simulate experimentally testable results. In this chapter, we examine, InsP$_3$-mediated calcium patterns in a neuroblastoma cell line, N1E-115, as modeled with and without spatial factors.

In the following studies, we examine hypotheses about the mechanisms required for InsP$_3$-mediated signaling in the neuroblastoma cell. These hypotheses are embodied in the computational model that is developed and constrained based on benchtop experiments. The results support the proposal that InsP$_3$ distributions and kinetics are determinants of calcium patterns in neuroblastoma cells. Fink *et al.* (1999b, 2000) hypothesized that temporal and spatial patterns of InsP$_3$ concentrations dictate the pattern of calcium release and subsequent temporal and spatial patterns of cytosolic calcium.

The researchers developed a quantitative model of the known and estimated characteristics of InsP$_3$ and calcium biology. These were used to test the hypothesis that enzyme kinetics and initial concentrations of the molecular components are sufficient to generate InsP$_3$ and calcium temporal dynamics. To explicitly explore the relationship between cellular morphology, diffusion, and calcium patterns, the computational model was modified to include spatial characteristics of the cell as an explicit spatial model. The initial models assumed proportional distribution of molecular species in the cell. A third hypothesized model developed by Fink includes experimentally determined spatial distributions of organelles and species that were found to be nonhomogenous (Fink *et al.*, 2000; Slepchenko *et al.*, 2003).

The noncomputational model that we build identifies the cellular processes involved in the biological behavior; defines the kinetics of the processes (rate equations); obtains parameter values for the components, concentrations, diffusion coefficients, rate constants for the rate equations; and defines topological features. The computational model implemented in Virtual Cell is validated by its ability to quantitatively reproduce experimental results and to predict experimental results that had yet to be performed. The two models we create in the next pages are slightly simplified versions of the ones published in Fink *et al.* (2000). We use simplified models such that we can focus on the construction of the biochemical reactions and the introduction of spatial features. Instead of creating the third model of Fink, we will discuss the model that can be found in the Virtual Cell database.

We will develop a noncomputational model for the intracellular mechanisms and reactions of calcium dynamics specific to the neuroblastoma cell type. This will include a set of system statements—the components and relationships between factors in terms of kinetic reactions. We then use Virtual Cell to implement the model and examine the role of spatial distributions on the simulation and understanding of calcium patterns.

7.3.1 Calcium Biology in the Neuroblastoma Cell

Cultured neuroblastoma cells that are bathed in saturating concentrations of the nonapeptide bradykinin, initiate a calcium wave in the center of the neurite that moves as a single bidirectional wave toward the soma and distal neurite (Fig. 7.4). The calcium release is mediated by activation of $InsP_3$ signaling and release of calcium from ER stores. Two to three seconds after bradykinin stimulation, calcium concentrations, measured from any location in the cell by fluorescence ratios, spike from 50 nm to 1000 nm in 1 second. The spike is followed by a gradual decrease in concentration.

Because we are specifically modeling $InsP_3$-mediated calcium dynamics within the neuroblastoma, our model is based on the biology of this cell (Fig. 7.5). The presence of $InsP_3$ and intracellular calcium release machinery ($InsP_3R$) in this cell type was identified by Surichamorn *et al.* (1990). $InsP_3$-mediated calcium release from intracellular stores occurs within the first 30 seconds after stimulation with bradykinin in N1E-115 cells. The maximal calcium release from internal stores occurs within the first 20 seconds. No measurable influx of calcium through calcium channels in the plasma membrane occurs within this time (Iredale *et al.*, 1992; Mathes and Thompson, 1994). By simulating the system within the first 20 seconds after stimulation, we monitor calcium release from intracellular stores and do not need to include calcium influx through the plasma membrane (PM). Intracellular stores of calcium in the neuroblastoma cell are located in the ER and mitochondria. Mitochondria do not contain $InsP_3$ receptors, and calcium release from these stores under bradykinin stimulation within the first 30 seconds has yet to be demonstrated. We will therefore leave them out of the model.

The effects of bradykinin stimulation on release of calcium from ER stores are attributed to $InsP_3$ actions (Reiser *et al.*, 1992; Ziche *et al.*, 1993; Mathes and Thompson, 1994). Therefore, we will model $InsP_3$ production but not model the specific mechanism of bradykinin stimulating $InsP_3$ production. The experimental system uses the calcium indicators as part of the experimental method to measure calcium levels. Because these act as biological buffers that affect the concentration and dynamics of free calcium, we will include the calcium indicator, fura-2, in the model.

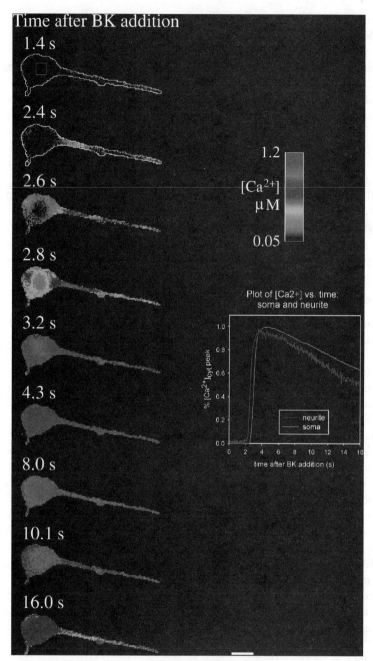

Figure 7.4 *A single NIE-115 cell loaded with Fura-2 and treated with 500 nm BDK to visualize changes in concentrations of cytosolic calcium. Images were captured with confocal microscopy at second time intervals; representative images shown over course of 12 seconds. Concentrations of calcium were measured in two locations: midneurite and soma. The insert shows measured concentrations plotted as a function of time for both locations. Note the increase in calcium concentration first at the neurite (2.3 seconds) followed by the soma and distal end of the neurite (3.2 and 4.5 seconds, respectively). BDK stimulation resulted in a maximum concentration of 1 mM within 4 seconds at both locations and propagated as a single wave in both directions. (Reproduced with permission from Fink et al., Biophysical Journal 79:163–183, 2000.)*

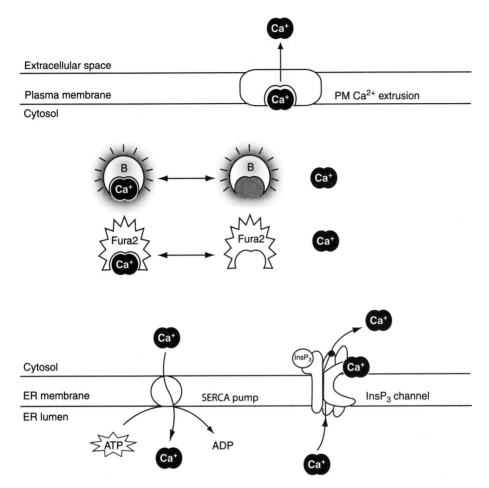

Figure 7.5 *Calcium mechanisms of the neuroblastoma cell modeled in this chapter. Calcium efflux is shown at the plasma membrane (PM) via a generalized plasma membrane pump. Endogenous (B) and exogenous (Fura-2) buffers bind to cytoplasmic calcium. Uptake and release of calcium from the endoplasmic reticulum (ER) is mediated by the SERCA pump and InsP$_3$ receptor channel, respectively. The influx of calcium through the plasma membrane is not included because it does not occur in the time frame of the simulation and is not included in the model.*

The NIE-115 cell line retains many of the morphologic and physiologic characteristics of *in vivo* neuronal cells (Richelson, 1979; Gill and Chueh, 1985). The nucleus is located within a large cell body, soma, and the neurite is a cytoplasmic extension, similar in shape to neuronal axons (Fig. 7.6). Geometrically, the neuroblastoma cell can be approximated with a hemisphere to represent the soma, a split cylinder for the growth cone, and a disk for the neurite (Fink, 1998).

7.3.2 System Statements

From our knowledge of the biological system, we can construct a set of statements that describe both its components and the relationships among them. These statements reflect what is known and accepted about the system based on the work of biologists

Figure 7.6 *Fura-2 filled neuroblastoma cell illustrating the cellular shape. The N1E-115 cell has structural and functional similarities to neuronal cells, specifically, identifiable soma and axon (neurite). (Reproduced with permission from Fink, Morgan, and Loew, Biophysical Journal 75, 1648–1658, 1998.)*

and biochemists. Although many complex statements can be made, simple statements facilitate creating the mathematical model.

Calcium interactions:

1. Free cytosolic calcium binds to buffer: endogenous, and exogenous (Fura-2).
2. Free cytosolic calcium is transported into the ER via SERCA pumps.
3. Calcium is released from the ER through Ins3P receptor.
4. Free cytosolic calcium is removed from the cell via extrusion mechanisms in the plasma membrane.

InsP$_3$ and InsP$_3$ receptor:

5. InsP$_3$ is produced near the plasma membrane and is degraded in the cytosol.
6. InsP$_3$ and calcium binding open the InsP$_3$ receptor when it is not inhibited.

Calcium Dynamics   The relationships identified in our statements indicate that calcium transitions from one state to another. Statement 1 indicates that free cytosolic calcium (Ca) binds to buffer and creates new calcium species, bound calcium (CaB, CaFura2). This is a binding reaction. Statement 2: free cytosolic calcium is also moved into the ER. The movement of a molecule from one physical compartment to another is referred to as a flux. The reactions and fluxes identified in the first two statements are reversible; bound calcium becomes unbound and ER calcium moves back into the cytosol. Statements 2 and 3 indicate the molecular species mediating the fluxes. Calcium translocation between the cytosol and ER is mediated by SERCA pumps (Statement 2) and InsP$_3$ receptor channels (Statement 3). These proteins participate in determining the rate of flow between calcium states.

Statement 4: Cytosolic calcium is removed from the cell via extrusion mechanisms in the plasma membrane. We know from our previous discussions that there are two dominant mechanisms for calcium to be extruded across the plasma membrane: Na/Ca exchangers and ATP pumps. The overall rate of extrusion affects cytosolic calcium dynamics. Because we are not focused on learning about the role and mechanics of calcium extrusion at the plasma membrane, it is sufficient to include that calcium is removed at a constant rate because of both molecular mechanisms. Therefore, the Na/Ca exchanger and ATP pumps are merged into a single process represented by a single flux rate (extrusion).

InsP$_3$ and Receptor Activation   Statement 5 indicates that concentrations of InsP$_3$ change over time due to its rate of production and degradation. The production of InsP$_3$ in this study is modeled by an immediate increase in InsP$_3$ concentration with decay in the production rate over time. InsP$_3$ is degraded by cleavage or phosphorylation in the cytosol (Wang *et al.*, 1995). The rate of that degradation is dependent on the concentrations of InsP$_3$ and a rate constant that reflects the joint activity of cleavage and phosphorylation.

Statement 6: InsP$_3$ receptors transition from inactive-closed channels to open channels when calcium and InsP$_3$ are appropriately bound (Ueda *et al.*, 1986; Bezprovanny, 1991). The InsP$_3$ channel consists of four peptides with three binding sites. Two of the three sites, one for calcium and the other for InsP$_3$, behave as cooperative activators of the channel. The third site, a calcium binding site, inhibits the channel. When calcium is bound to the inhibition site, the channel is unable to open regardless of binding at the activation sites. The fraction of receptors inhibited by calcium can be modeled as a probability based on calcium concentrations (Keizer and De Young, 1992; Li and Rinzel, 1994; Li *et al.*, 1995).

7.3.3 What Are the Kinetics?

The overall concentration of calcium within the cytosol is defined by the set of processes that involve calcium binding and translocation from one cellular location to another. This can be written as the following sentence: The concentration of calcium over time is a function of the rate of release from the ER stores (input) and the rate of cytosolic calcium binding to buffer, the rate of extrusion of through the plasma membrane, and the rate of uptake into the ER (outputs). This statement can be rewritten as an ordinary differential equation:

$$\frac{d[\text{Ca}_{\text{cyt}}]}{dt} = \text{rate of release} - (\text{rate of buffering} + \text{rate of extrusion}$$

$$+ \text{rate of ER uptake})$$

Similarly, the change in concentration of InsP$_3$ can be stated in a sentence as: The change in concentration of InsP$_3$ over time is determined by the rate of its production minus the rate of its degradation. The differential equation then can be written as follows:

$$\frac{d[\text{InsP}_3]}{dt} = \text{rate of production} - \text{rate of degradation}$$

Each of the processes in the differential equation can be described as a rate equation. The rate equation describes the speed at which the process occurs given the concentration of the species and the molecular mechanism of the process. Importantly, each reaction rate is its own hypothesis on the mechanics of the process. As such, each kinetic rate equation has its own assumptions or rationale. We walk through each reaction in the model to make the choice of kinetic reactions visible. The rationale for each kinetic type may have been established in previous biochemical research or may be an estimated assumption. The next section is useful for obtaining an understanding of the basis for the kinetics. When one creates models, one will provide one's own set of justifications, assumptions, or rationale for why a particular kinetic model is appropriate for the cellular process. It is possible to skip this section and look at the composite table of equations and parameter values provided in the Appendix. The composite table is similar to many differential equation and parameter tables provided in modeling literature.

TABLE 7.1   Rate of Ca Binding and Release of Buffer

Rate equation	$k_{on}[Ca][B] - k_{off}[CaB]$
where $K_{eq} = \dfrac{k_{off}}{k_{on}}$	

7.3.3.1   Calcium Kinetics

BINDING BUFFERS   The reaction between cytosolic calcium (Ca) and endogenous and exogenous buffers (B, Fura2) to produce bound calcium (CaB, CaFura2) is reversible and can be modeled as mass action kinetics (Table 7.1). The numerical value for these rates are determined by the relationship to the equilibrium constant K_{eq}.

FLUXES   The uptake of calcium from the cytosol to the ER through the SERCA pumps, release of calcium into the cytosol through the InsP$_3$ receptor, and extrusion from the cytosol to the extracellular space are all fluxes. A flux by definition is the movement of mass across an area. This describes the movement of molecular species across the membrane. Fluxes, like reactions, can be described by different kinetic types.

CALCIUM EXTRUSION AT PM   The extrusion of calcium from the cytosol is mediated by pumps whose dynamics can be modeled here as general mass flux. General mass flux describes the unidirectional movement of a molecule through a membrane's surface area as proportional to the transported molecule's concentration. The pumps are represented by a rate constant (k) that is derived from their combined average peak flow rate. The rate of transport in our model is also dependent on the concentration of calcium being above a threshold (0.2) and the amount of calcium above the threshold. These are also taken into account in the rate equation (Table 7.2).

CALCIUM INFLUX TO ER: SERCA KINETICS   Flows mediated by InsP$_3$ receptors and SERCA pumps may be best described by more complex equations than mass flux. The activity of the SERCA pumps and InsP$_3$ receptors are best described by Hill-type kinetics that take into account cooperative affects of calcium on the rate of uptake by the pump (Gill and Chueh, 1985; Lytton *et al.*, 1992). Generically, Hill-type kinetics follow the form:

$$V_{max} \frac{[S]^n}{[S]^n + K_d^n}$$

where S is the substrate concentration, K_d the dissociation constant, and n the cooperativity coefficient. For the SERCA pump, in relation to calcium, we can write the reaction rate as in Table 7.3. J_{max} is used in place of V_{max} to indicate the flow across a membrane.

CALCIUM RELEASE FROM ER: InsP$_3$R KINETICS   InsP$_3$ receptor peptides have three binding sites: one calcium site that inhibits channel function and two other sites (one calcium,

TABLE 7.2   Rate of Ca^{2+} Extrusion Cytosol to Extracellular Space

Rate equation	$k([Ca] - Ca\ Threshold)([Ca] > Threshold)$

TABLE 7.3 Rate of ER Update of Ca^{2+}

Rate equation	$J_{max}\dfrac{[Ca]^2}{[Ca]^2+K_d^2}$

TABLE 7.4 Rate of ER Release of Ca^{2+}

Rate equation	$J_{max}\left[\left(\dfrac{[InsP_3]}{[InsP_3]+K_{InsP_3}}\right)\left(\dfrac{[Ca]}{[Ca]+K_{act}}\right)\right]^3\left(1-\dfrac{[Ca]}{[Ca_{ER}]}\right)$

one InsP$_3$) that cooperatively stimulate function. To include calcium and InsP$_3$ binding as factors in the activation of InsP$_3$ receptors, we must explicitly include them in the kinetic rate equation of the flux. The role of calcium and InsP$_3$ as activators of the InsP$_3$ receptor can be modeled with a Hill coefficient for the two regulators.

The kinetic model in Table 7.4 describes calcium and InsP$_3$ in terms of activators. There is no representation of channels inhibited by calcium, which are an important factor in the flow of calcium. Calcium inhibition of the InsP$_3$ receptor can be modeled as a probability factor, h (Fink *et al.*, 2000) The factor h is determined by the concentration of calcium, the rate of calcium binding to (k_{on}) and release from (k_{off}) the inhibition site. Therefore, the numerical value of h at any given time is a function of the concentration of cytosolic calcium (Ca) and the dissociation constant of calcium to the InsP$_3$R inhibition site. This is described as a reaction between Ca and h in Table 7.5.

To reflect the affect of calcium inhibition on the rate of channel opening, we modify the previous Hill coefficient reaction (Table 7.5) to include the probability factor, h (Table 7.6).

7.3.3.2 InsP$_3$ Reaction Kinetics

The rate of InsP$_3$ production in response to brady-kinin stimulation was unknown and needed to be estimated. It was known that the rate of InsP$_3$ production decays over time. Therefore, it was decided to model InsP$_3$ production as an initial rate (J_{InsP_3}) with a decay function similar to radioactive decay (e^{-kt}). Degradation of InsP$_3$ also occurs in the cytoplasm. The kinetics of this degradation will be modeled as mass action with the degradation rate constant (Table 7.7).

TABLE 7.5 Probability of Inhibition Site Being Occupied

Rate equation	$k_{on}(K_{in}-([Ca_{cyt}]+K_{in})*h)$

TABLE 7.6 Rate of ER Release of Ca^{2+}

Rate equation	$J_{max}\left[\left(\dfrac{[InsP_3]}{[InsP_3]+K_{InsP_3}}\right)\left(\dfrac{[Ca]}{[Ca]+K_{act}}\right)h\right]^3\left(1-\dfrac{[Ca]}{[Ca_{ER}]}\right)$

TABLE 7.7 Rate of Production and Degradation of InsP$_3$

Rate equation	$J_{InsP_3}e^{-kt}$
Rate equation	$k(InsP_{3cyt}-InsP_{3init})$

7.3.4 Inclusion of Geometry and Diffusion

At this point in the chapter, we have a conceptual diagram of the biology, generated a set of system statements, and have written rate equations for the kinetics of each process. Together these define a continuous, open-system model of calcium dynamics that consists of populations of molecular species that change over time. Despite describing the kinetics of the calcium release, extrusion, and uptake, there is yet no explicit specification of space or movement of molecules within the model. The movement of the molecules is conceptual. To explicitly take into account the movement of molecules within the cell, we need to include the spatial properties of the cell. The spatial properties include descriptions of the cellular geometry, molecular location, and movement.

GEOMETRY The shape of a cell can be described on a computer analytically, that is, sphere, rectangle, hemisphere, or extracted from cell imagery. Tracing the surface of cells within a single image generates a mathematical representation of the 2D cell shape. A 3D version can be generated by tracing across a Z-series or stack of images. A common approach to solving reaction equations within the geometry is to generate a mesh. The mesh is a grid of x,y (2D) or x,y,z (3D) coordinates that define the entire geometry. Reaction equations for the biochemical processes are subsequently solved within each mesh cell. Molecules are assumed to be uniformly distributed within each cell.

LOCATION The cellular structures created in the concept map of the physiologic model correspond with physical compartments within the cell. As such, cellular locations are a compartment with characteristics of area and volume that contains species and reactions as well as an object that has its own density distribution within the cell. The distribution of molecules and organelles in space can be described as a density: amount/volume. When the distributions are homogenous, the density factor for all components is set to 1 (i.e., uniform). To vary the density of an organelle or molecular species, we need only change the density factor. To take density into account as a factor, species, we would add it both to the physiologic model and the rate equations of membrane reactions and fluxes as a multiplier (see Appendix). We will see that Fink *et al.* (2000) did this for their third model, which takes into account nonuniform distributions.

MOVEMENT Once the spatial model has been generated, diffusion of species can be modeled by inclusion of diffusion coefficients that define the kinetics of how the species move. Models of reactions that take into account diffusion are classified as reaction-diffusion systems. Mathematically, reaction-diffusion systems require solving partial differential equations. To write the differential equations ourselves, we would include a term for the diffusion of soluble factors. For example, the initial differential equations for changes in calcium concentrations over time would be rewritten to the following:

$$\frac{\partial[Ca]}{\partial t} = D\nabla^2[Ca] + \text{rate of release} - (\text{rate of buffering}$$

$$+ \text{ rate of extrusion} + \text{rate of ER uptake})$$

The tool we have chosen to work with, Virtual Cell, creates the differential equations for us. This makes it only necessary for us to provide the diffusion coefficient for the diffusing molecules.

7.3.5 What Are the Initial Conditions and Parameters?

From our conceptual model and set of system statements, we can create a legend of variables and their associated parameter values. We use values derived from experimental data to set the initial conditions of the variables. The more experimental data we have on concentrations and binding constants, the more constrained and presumably better the representation of the biological system will be in the model. For each variable, we write what we know about its concentration and diffusion coefficients.

The pumps and channels that facilitate the movement of calcium through the cell have characteristic maximum rates of transport in addition to binding constants for the carrier molecule or ones by which they are regulated. Ideally, all of the parameters come from the same biological system and experimental results. The numerical value for the initial rate of InsP$_3$ production (J_{InsP_3}) upon stimulation and the rate constant for InsP$_3$ decay (k) were unknown parameter values. The authors therefore explored a range of parameter values to determine which combination would reproduce the obtained experimental values for cellular concentrations of InsP$_3$ after bradykinin stimulation (Fink *et al.*, 2000).

Modelers turn to the literature to draw on what biophysicists and biochemists have already determined about how channels and pumps function (Gill and Chueh, 1985; Lytton, 1992; Wang *et al.*, 1995; Herrington *et al.*, 1996; Miyawaki *et al.*, 1997; Meldolesi and Pozzan, 1998). We will use the same values reported by Fink *et al.* (2000) (Table 7.8; Appendix). The complete set of parameter values, initial conditions, and equations have been combined in the Appendix as a single reference for creating the numerical model within Virtual Cell.

TABLE 7.8   Initial Condition and Parameter Values of Model

Molecular Species	Initial Conc. (μM)	Diffusion Coeff.
InsP$_3$_init[b]	0.16	283
Ca^{2+}		
External	2000	
ER[c]	400	
Cytosol[b]	0.05	220
Buffer		
Total exogenous (Fura-2)[a,b]	75.0	50
Total endogenous[a,b]	450	
Cytosol[b]	0.05	
Probability (*h*)	0.8	
Fixed species	1 (no unit)	
InsP$_3$R	1	
SERCA	1	

[a]The total amounts of buffers are distributed between bound and unbound states at equilibrium. The equilibrium values are used for the initial concentrations; see Appendix.
[b]Fink *et al.* (2000).
[c]Miyawaki *et al.* (1997); Meldolesi and Pozzan (1998).

7.4 MODELING CALCIUM DYNAMICS WITH VIRTUAL CELL

We have identified all aspects of our mathematical model: factors, reactions, rate equations, and parameter values. To complete this as a computational model, we enter the model features into a computational framework that can create and solve the necessary ordinary and partial differential equations. In this chapter, we use Virtual Cell to develop, based on our model features, the appropriate ordinary and partial differential equations needed to model temporal and spatial characteristics. We will not explore in the context of this chapter the details of solving partial differential equations, but rather note that the inclusion of space creates a computational challenge requiring modeling software that can solve such equations. Virtual Cell is designed to represent cellular systems. It enables cell biologists to use familiar terminology, concepts, and imagery to develop complex models of cellular systems.

Detail 7.1

Virtual Cell is a simulation tool that runs from a remote server. It is accessible via Java applets that enable one to access and run the software from any computer with the appropriate Java runtime installed (currently JDK 1.3) and connected to the Internet. In order to use the tool, one must create a user login and password. Login instructions are present at the site (http://vcell.org/login/login.html).

The case study we are about to complete with Virtual Cell compares three simulations of calcium dynamics. The first model is a simulation of a compartmental model that assumes a single, well-mixed pool of reactions that does not take space into account. The second model applies the same set of reactions to the spatial dimensions of the neuroblastoma cell, and the third takes into account the nonuniform distribution of cellular machinery within the cell. This parallels the work of Fink *et al.* (2000) and will allow us to see how the inclusion of spatial features can affect our understanding of the mechanisms involved in cellular calcium dynamics. The text below describes the creation of the first two models. The results of the third model are taken from the original model, which is stored as a shared model in the Virtual Cell (http://vcell.org).

7.4.1 Using Virtual Cell

Virtual Cell has three main databases: BioModel, MathModel, and Geometry. We will be creating a BioModel that will be stored in the BioModel database and accessing images within the Geometry database that will be associated with the BioModel. A Bio-Model consists of a physiologic model, applications, and simulation results (Slepchenko *et al.*, 2003). The physiologic model contains the icon and math-based description of the cellular structures, membranes and cellular compartments, molecular species, and reactions and fluxes. Applications contain the experimental conditions for the model simulation including spatial models and simulation results. The Geometry database stores personal and shared image files. The created or imported image files define the spatial topologies in which reactions occur, making spatial models. These can be analytic 1D, 2D, or 3D shapes—spheres, circles, squares—or images of cell outlines. MathModels are for users who want to create their models by directly writing the mathematical equations and initial conditions. The published models of Fink were written as MathModels. As such, the authors could directly simplify equations and combine terms within the differential equations. The differences between BioModel and the original MathModels developed by the authors lay within the representation of parameter values and conversion factors. In the following pages, we use the BioModel interface and an image stored in the Geometry database to create our computational models.

7.4.2 BioModel Workspace

Virtual Cell opens to the BioModel workspace where we create the "Physiology" of the model. Here we use icons to add and name the cellular structures and molecular species. Adding a cellular compartment creates a membrane, interior and exterior location. Any features added to a BioModel must be named. Names are assigned to the membrane and the interior compartment. For our model, we add one compartment for the cytosol with plasma membrane (PM) and add two more circles within the first compartment to create the ER with ER membrane and the nucleus with nuclear envelope (Fig. 7.7).

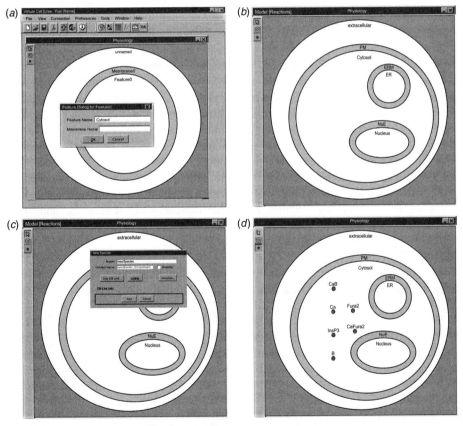

Figure 7.7 (a) Virtual Cell physiology workspace showing the dialogue box for creating cellular structures. (b) Cellular structures included in our model are the plasma membrane (PM), cytosol; nuclear envelope (NuE), nucleus; endoplasmic reticular membrane (ERM), endoplasmic reticulum (ER). (c) Dialogue box for creating species. Species are named and the context name is generated. (d) Cytosolic species included in our model are shown.

Detail 7.2

Creating Cellular Compartments

Place a circle for the cellular compartment in the workspace by selecting the membrane icon and clicking once within the workspace. A name box appears with fields for the "Feature Name" of the interior space, cytosol, and the "membrane" name, PM (Fig. 7.7). The process is repeated to create the ER, nucleus, and their associated membranes.

Create one compartment for the cell (cytosol, PM)

Create two more compartments within the cytosol

  One for nucleus, nuclear envelope (NE)

  One for ER, ER membrane (ERM)

7.4.2.1 Species

We add molecular species, components, to the cellular compartments and structures. Icons of species serve as conceptual aids for discussion, notation for what is in the model and factors in our reactions and fluxes. Species are named as they are added to the model workspace. The common name one gives the species is used by Virtual Cell to generate a context name (Table 7.9). The "context name" combines both the common name and the name of the compartment in which the species is located. The contextual name uniquely identifies a name for the species state. The "formal name" is the name by which one may find this molecule in literature. The "common name" is how the molecule will be identified in the model. DBlink allows one to identify the species as equivalent to other species that already exist in the Virtual Cell database. Although not required in order to create a model, it does make it possible for one to search the Virtual Cell database for public models using the same species.

We add species for each component of the model in their respective cellular locations. In the cytosolic compartment we add species for calcium (Ca), endogenous buffer (B), exogenous buffer (Fura-2), calcium bound to buffers (CaB, CaFura2), and $InsP_3$. Calcium is also added to the ER and extracellular space. The SERCA Pump, $InsP_3$ receptor channel, and probability are added to the ER membrane.

TABLE 7.9   Species Names in Virtual Cell

Formal Name	Given Name	Generated by Virtual Cell	Location
Calcium	Ca	Ca_Cytosol	Cytosol
Buffer	B	B_Cytosol	Cytosol
Bound calcium	CaB	CaB_Cytosol	Cytosol
Fura-2	Fura2	Fura2_Cytosol	Cytosol
Fura-2 bound	CaFura2	CaFura2_Cytosol	Cytosol
Inositol triphosphate	$InsP_3$	$InsP_3$_Cytosol	Cytosol
Calcium	Ca	Ca_ER	ER
SERCA	SERCA	SERCA_ERmembrane	ERmembrane
$InsP_3R$	$InsP_3R$	$InsP_3R$_ERmembrane	ERmembrane
Probability	*h*	h_ERmembrane	ERmembrane

Note: Naming conventions used in original papers have been retained for clarity in comparison of chapter to revised paper.

Detail 7.3

Adding Species to the Physiologic Model, Calcium Example

To create calcium in the cytosol, select the species icon ⊙ and click inside the cytosol. Enter a name for the species in the dialogue box that appears (Fig. 7.8). For example, the formal name of calcium is calcium. Its common name is Ca. The context name created by Virtual Cell will be Ca_Cytosol.

To add the same species to multiple compartments: Right click to open the edit menu. Select copy. Move to the desired compartment. Right click for the edit menu and select paste. Virtual Cell creates the appropriate context name for the species. Calcium in our model is added to three compartments (external, cytosol, and ER).

7.4.2.2 *Reactions and Fluxes* Reactions between species or fluxes that move species across membranes between compartments are features of each cellular structure. "Reactions" are characteristic of compartments, for example, cytosol or membranes (e.g., plasma membrane), and "Fluxes" are specific to membranes. Reactions and fluxes

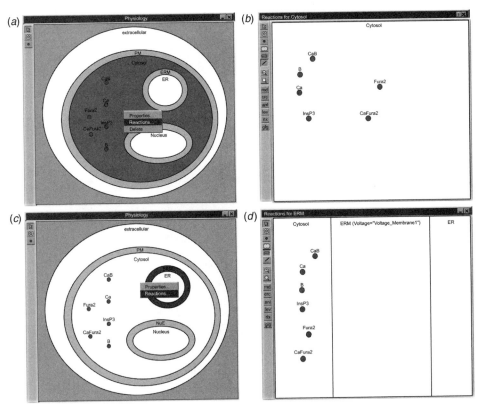

Figure 7.8 *The kinetics of reactions and fluxes are created using Reaction editors. Reactions and fluxes are properties of the cellular structures. (a) and (c) Accessing reaction and flux editors of cellular structures, cytoplasm, and ERM, respectively. (b) The reaction workspaces for cellular compartments. (d) The workspace for membranes. (b) and (d) Species added in the Physiology workspace are visible in the windows.*

are created within reaction editors of the selected compartment (Fig. 7.8). These editors are titled "Reactions for *structurename*." The editors differ in appearance for compartments and membranes. However, the process for creating reactions and fluxes are nearly identical. Both are associated with a cellular structure, given names, and linked to species and assigned kinetic types.

Reactions   The reactions are created and edited through the Reaction editor "Reactions for *structurename*," (e.g., "Reactions for Cytosol" Fig. 7.9). The reaction editor is used to assign a name, define participating species, and assign kinetics. In the "Reactions for Cytosol" window, we find a tool pallet for constructing reactions. Reactions are created by linking participating species to a reaction icon. For example, we draw lines between the reaction icon for endogenous buffering, the reactants (Ca_Cytosol, B_Cytosol) and product (CaB_Cytosol) to create the binding reaction of calcium to endogenous buffer. Repeating this process for all reactions in the appropriate compartments (Table 7.10) completes defining the components and relationships of our physiologic model.

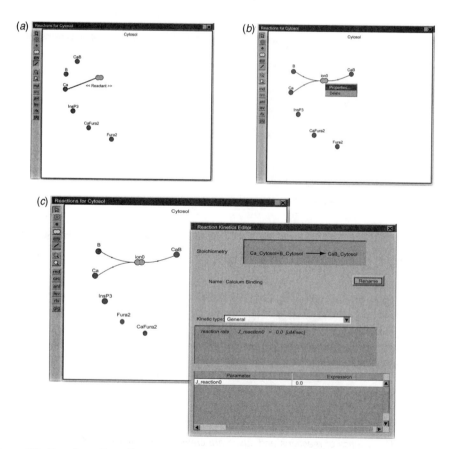

Figure 7.9   *Reaction editors. Reactions are defined by double-clicking the "reaction" icon to access the Reaction editor. Select the kinetic type and enter the appropriate rate equation in the expression field. See Appendix or tables in the chapter for equations.*

TABLE 7.10   Location of Reactions and Fluxes in Cellular Structures

Reaction/Flux	Cellular Structure
Calcium extrusion	Plasma membrane
InsP$_3$ generation	
Calcium binding to Fura-2	Cytosol
Calcium binding to endogenous buffer	
InsP$_3$ degradation	
Probability of occupied InsP$_3$ calcium inhibition site	
SERCA uptake of calcium	ER membrane
InsP$_3$ receptor release of calcium	

Detail 7.4

Adding Reactions and Reaction Kinetics to Compartments

Select the structure in which the reaction takes place. Right click to access the dialogue box containing "Reactions" (Fig. 7.8). Select "Reactions." A window titled "Reactions for *structurename*" (i.e., "Reactions for cytosol") will open.

Select the bar-bell styled reaction icon ⬭. Add it to the compartment.

Next link the reactants and products to the "Reaction" icon to signify those involved in the reaction, that is, product (CaB_Cytosol), reactants (Ca_Cytosol, B_Cytosol), or catalyst. The relationships of species to reactions are defined by drawing lines from the species to the reaction icon. Three names appear as the line is drawn: "reactant," "product," and "catalyst." These categories define the role of the species in the chemical reaction. Reaction editors are used to define the rate equation for each reaction. Double click the "Reaction" icon to access the Reaction editor.

The chemical reaction is shown in the editor as a schematic with substrates (left of arrow), catalysts (above arrow), and products (right of arrow). From the Kinetic Type menu, select the type of kinetics that governs the reaction. The rate equation is entered into the "Expression" field and the parameters (e.g., K_m, v_{max}, etc.) of the equation appear as a list. Any new symbol or text character(s) entered in the expression field is treated as a new parameter in the reaction and added to the parameter list.

To define the kinetics of the interactions, we use the Reaction Kinetic Editor to assign rate equations to each reaction we created. Reaction Kinetic Editors contain a chemical reaction diagram with reactants to the left (Ca_Cytosol, B_Cytosol), products (CaB_Cytosol) to the right (Fig. 7.9), and any catalyst above the arrow. "Catalyst" is a species that participates in the reaction but is neither substrate nor product. Enzymes, pumps, and channels are most often catalysts. We choose the type of kinetic reaction in the reaction editor. The kinetic type tells the software how to relate mathematically to the reactants, products, and catalysts in the reaction. The reaction drawn in the "Reactions of Compartment" editor and the selected kinetic type are used to identify variables and parameters. The kinetic type for the reactions in our model is "General." The rate equations we defined previously for each reaction are written into the "Expression" field. The syntax used in the expression field is similar to writing equations within Excel. The syntax for each rate expression we use has been provided in the Appendix.

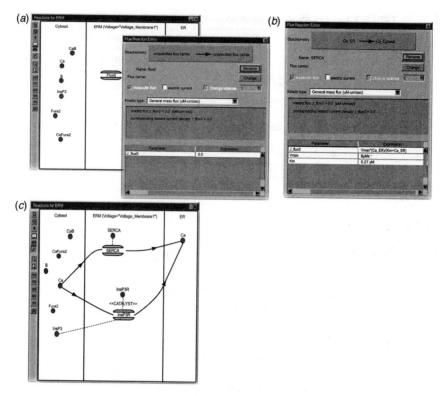

Figure 7.10 *Flux editors like Reaction editors are accessed by double-clicking the "Flux" icon. Transported molecules (calcium) and rate equation are entered in this window. Catalysts are added by drawing linkages to the "Flux" icon after the flux is created in the Reaction editor.*

Flux Creating fluxes is generally identical to creating reactions: add flux icon, name flux, identify participating species, and define kinetics (Fig. 7.10). There are two features that become obvious while working with the flux tool. In reactions, we use line tools to identify participating species as substrate, product, or catalyst. In fluxes, drawing a line between species and the flux icon defines the species only as a catalyst. Transported species are specified through the flux editor, and the lines are generated for us. It also becomes noticeable that the arrows represent a flow through the membrane, but not the directionality of the flow. The actual direction of the flow is determined by the rate equation as it is written in the rate expression. At the plasma membrane and ER, calcium is moved out of compartments. If we were drawing this by hand, we would probably draw the arrows from right to left to indicate directionality. In Virtual Cell, they are drawn left to right, and the actual direction of the flow depends on the balance between positive and negative terms in the equation. This is true for both fluxes and reactions.

Detail 7.5

Adding Fluxes to Membranes

Fluxes are added to membranes in a similar manner as reactions are added to compartments.

Select the PM membrane, right click and select "Reactions" from the menu that appears (Fig. 7.10). A window opens that shows three columns. Each column represents a compartment for the external space (left), membrane (middle), and internal compartment (right). Any species that have already been added to the model are visible in their respective compartments. To add a reaction, select the "Reaction" icon, add it to the center field, and define the reaction by linking species.

To add a flux, select the flux icon ⊜ and click in the middle column to add it to the membrane. The Flux editor dialogue box opens immediately.

Click the "Rename" dialogue box and enter the name of the flux (i.e., extrusion or SERCA). The "flux carrier" is the species to be transported. Change the flux carrier to the appropriate species (e.g., Ca). We leave electric current unchecked because we are not examining the effects of current on the system. Select the Kinetic Type for the flux (e.g., General Mass Flux).

7.4.3 Application Pallet

The physiologic model created in the BioModel Workspace identifies variables, parameters, and reactions of the system based on diagrams familiar to biologists. In order for the model and simulation to be complete, we must relate these features, equations, and dynamics to parameter values and a spatial description. The Application Pallet is the tool used to put the physiologic model together with initial conditions, active reactions, and any selected geometry. The Application Pallet contains five function tabs: "Structure

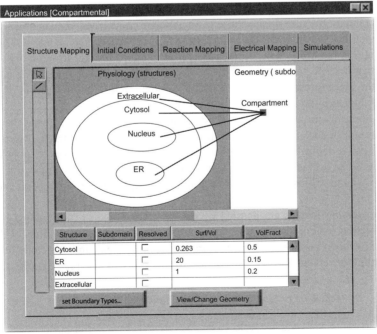

Figure 7.11 *Application Pallet interface provides tabs for completing the computational implementation: "Structural Mapping," "Initial Conditions," "Reaction Mapping," "Electrical Mapping," "Simulations." The Application Pallet opens to the "Structual Mapping" window.*

Mapping," "Reaction Mapping," "Initial Conditions," "Electrical Mapping," and "Simulations" (Fig. 7.11). Because we are not working with membrane potentials or other electrochemical reactions, we will not use "Electrical Mapping." Each application, "Compartmental" and "Uniform," we create in this chapter, differs in its configuration of the structure map or initial conditions. Each uses the same physiologic model.

Detail 7.6

Application Pallet: New Applications

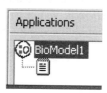

In the Application window, select the BioModel icon. Right click to access the "Create New Application" menu. Name the application "Compartmental."

7.4.3.1 *Structure Mapping*

The "Structure Mapping" tool is the user interface for applying the physiologic model to selected geometries. The diagram of the physiologic model is shown in the Geometry editor and used to map relationships between cellular structures in the diagram and geometry. By default, the model is mapped to a single point (Fig. 7.11). The single point is a diagram representation of the compartmental model that assumes the physiologic model reactions occur in a single, well-mixed compartment.

Each cellular structure created in our physiologic model, whether mapped to a geometry or not, has two properties: *surface volume ratio* and *volume fraction*. The surface volume ratio is the ratio of the surface area to volume of a compartment. We know for instance that the ER has a high surface-to-volume ratio because of its intricate folding, whereas the cell itself has a lower ratio. The volume fraction is the fraction of total cell volume occupied by the structure (Table 7.11).

The volume fractions are important for compartmental models because the units for concentrations in the simulations are mass/volume. Fluxes are defined as $\mu M/\mu m^2$, which are converted to micromolar concentrations as the molecules enter nonmembrane structures. When species move from one structure to another, the volume fractions and surface-to-volume ratios are used to determine the appropriate conversion factors for the reaction.

The conversion factors are defined by the software based on the volume ratios and volume fractions entered by the user. They are applied to the rate equations of the membrane fluxes where molecules move from one compartment and structure to another. Because the conversion factors are applied to all terms of the rate equations, to keep

TABLE 7.11   Spatial Ratios of Cellular Compartments

Structure	Surface Volume	Volume Fraction
Extracellular		
Cytosol	0.263	0.5
ER	20	0.15
Nucleus	1	0.2

the same rate constants for the enzyme kinetics found experimentally and used in the original MathModel, the numerical value of the rate constants must be recalculated to retain the experimentally defined values.

Highlight 7.1

Flux Conversions

The flux conversion for a membrane to the volume it defines (e.g., ERM to ER or NE to nucleus) is the surface-to-volume ratio (SurfToVol) (Eq. 7.1.1). When a cellular volume (e.g., Cytosol) contains subdomains, ER and Nucleus, the volume fractions of the subdomains are also taken into account (Eq. 7.1.2).

$$KFlux\_ERM\_ER = SurfToVol\_ERM \qquad (7.1.1)$$

$$KFlux\_PM\_Cytosol = \frac{SurfToVol\_PM}{1.0 - VolFract\_Nucleus - VolFract\_ER} \qquad (7.1.2)$$

$$KFlux\_ERM\_Cytosol = SurfToVol\_ERM * \frac{VolFract\_ERM}{1.0 - VolFract} \qquad (7.1.3)$$

Example: BioModel Conversion for SERCA Flux

BioModel Conversion: Rate Equation*Kflux_ERM_Cytosol

$$KFlux\_ERM\_Cytosol = SurfToVol\_ERM * \frac{VolFract\_ERM}{1.0 - VolFract}$$

$$SurfToVol\_ERM = 20 \qquad VolFract\_ERM = 0.15 \qquad 1.0 - VolFract = 0.85$$

$$Kflux\_ERM\_Cytosol = 3.529412$$

Rate equation: $J_{max} \dfrac{[Ca_{cyt}]^2}{[Ca_{cyt}]^2 + K_p^2}$

Converted rate equation: $J_{max} \dfrac{[Ca_{cyt}]^2}{[Ca_{cyt}]^2 + K_p^2} * Kflux\_ERM\_Cytosol$

$J_{max} = 3500\,\mu Ms^{-1} Kflux\_ERM\_Cytosol = 3.529412$

Converted rate equation with adjusted J_{max}: $991.666 \dfrac{[Ca_{cyt}]^2}{[Ca_{cyt}]^2 + K_p^2}$

7.4.3.2 Reaction Mapping The "Reaction Mapping" window provides a list of all the reactions and fluxes created in the model with options to turn reactions on or off ("Enable") or to make them very fast ("Fast"). When "Fast" is selected, we indicate that the reaction is near equilibrium and that it occurs on a scale faster than its rate of diffusion such as when diffusion coefficients are part of the reaction equations. For this set of simulations, we make the calcium-buffer reactions fast (see Chapter 4).

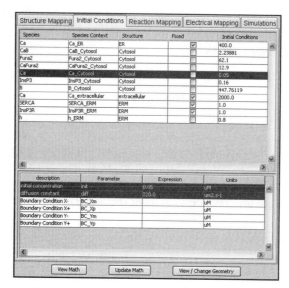

Figure 7.12   *"Initial Conditions" tab of the Application Pallet lists all variables of the system. In spatial models, the diffusion coefficient for soluble molecules is entered in the lower panel. Concentrations for species that do not change throughout the simulation can be set to a fixed concentration value in this window as seen for the transport proteins in the membrane and ER calcium concentration.*

7.4.3.3 Initial Conditions

Under the "Initial Conditions" (IC) tab, we see a list of the species that we added to each compartment. Initial conditions are saved as part of the application and therefore independent of the physiologic model. This allows us to set up multiple scenarios without having to redraw the basic topology of the system. We specify initial concentrations for each species and set it as fixed or variable during the simulation (Fig. 7.12). Within our model, membrane proteins are fixed, and the rest are variable. We can make a copy of the complete compartmental model so that we use the same IC values in our spatial model, "Uniform."

7.4.3.4 Spatial Models

Adding Geometries   Spatial models are created by solving the biochemical reactions within defined geometric shapes. Currently, the only identified geometry is the compartmental model, which has descriptions of volume fractions and surface area but no explicit spatial features (Fig. 7.13). It is the well-mixed homogenous compartment assumed for all ODEs and most enzyme kinetic models. Geometries are added to the Application Pallet through the "View/Change Geometries" tab. We change the geometry by selecting the desired image files from the Geometry database.

Detail 7.7

Adding Geometries: Spatial Models

Highlight the Compartmental application. Right click to access the command menu and select "Copy." Name the new application: "Uniform." In the new application pallet,

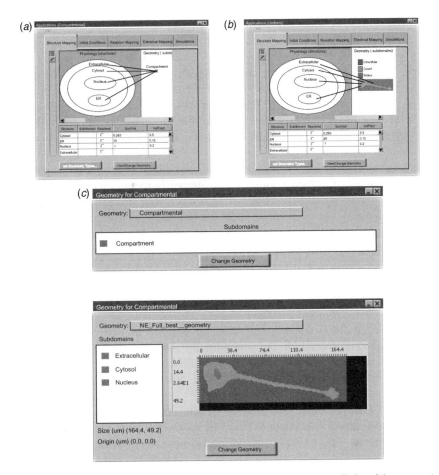

Figure 7.13 *Changing geometries and duplicating models. The Application Pallet of the compartmental model is shown at left (a), the uniform distribution model at right (b). The "View Change" icon at the base of the first two windows (a, b) is used to access the Geometry dialogue boxes (c). Once a geometry is selected, it is added to the application.*

click the "View/Change Geometry" tab. Select "Change Geometry." This accesses the Geometry Database and "Shared Geometry" folder. The geometry we will use is in the folder "CellBioGuide." Highlight and select the image file "NE-Full-Best_Geometry." This will update the structure mapping view with the geometry.

Resolving Structures

Use the line tool provided in the Application Pallet to draw a line from the "Physiology" model structures to the squares that correspond to regions in the cell geometry. As the line is drawn between the two images, the word "RESOLVED" will appear to indicate that Virtual Cell has mapped the model compartment to the region in the cellular geometry file. We do not resolve membranes in the models of this chapter.

The Geometry database view contains one's image folders as well as folders of images made publicly available by other users. A single image may be applied to multiple BioModels. A model of RNA transport can be applied to the neuroblastoma cell geometry as easily as the calcium dynamics model. Similarly, we can apply two different geometries to the same physiologic model. Instead of examining calcium dynamics in the neuroblastoma, we could examine them in oocytes that have a completely different geometry. The analysis of Fink *et al.* (2000) was done with a 2D image of the neuroblastoma cell. The image associated with Fink's publication is stored in the geometry database as a public file that can be imported into one's own workspace. Each image file has information describing its dimensions and size that is imported with the image.

Once the geometry is linked to the application, the cellular structures of the physiologic model must be mapped to their corresponding regions in the image. This allows Virtual Cell to map the volumes in the BioModel to the newly linked area and volume data associated with the image. Once these are mapped, the structures are said to be "resolved," and the volume fractions that we created in the compartmental model are replaced by the Virtual Cell calculated volumes.

Diffusion To describe how molecules move through the spatial model, we assign diffusion coefficients. Diffusion coefficients are a characteristic of individual molecular species in their structural context and therefore assigned as part of the "Initial Conditions" (Fig. 7.12). Not all of our variables diffuse; therefore a default setting of 0.0 for these molecules is appropriate. The diffusion values used in our spatial model, "Uniform," for InsP$_3$, calcium, and fura2 are provided in Table 7.8.

7.4.3.5 Simulations The final step of creating an application is to set the simulation conditions. In the "Simulations" tab, we instruct Virtual Cell how to solve the mathematical model derived from the information already provided through the construction of the "Physiology" model, the initial conditions, and geometries. This is done by choosing an approximation method for the model, selecting a number of time points to be sampled, and defining the resolution of the geometries mesh for spatial simulations. The time it takes for the simulation to compute is affected by the number of time steps that are taken and recorded, the resolution of meshes in spatial models, and the approximation method. The approximation method is selected by Virtual Cell based on the mathematical description generated for one's model. The advanced settings allow users to choose the integration method for solving the set of ODEs that describe the model. The compartmental simulations discussed in this chapter use the Runge-Kutta fifth order integration method. The approximation method for the spatial models is set by Virtual Cell.

Detail 7.8

Application: Simulations Control, Edit

The simulation tab allows you to create simulations and manage simulation results stored on the remote server. Click "New" to create a simulation. Click "Edit" to set the simulation methods.

Use the "Parameters," "Tasks," and "Advanced" tabs to edit the default value for parameters, time step, and integration method, respectively.

Parameters: Change the rate constant values for InsP3 production ($J_s = 17.73$), calcium extrusion ($v_{max} = 6.08$), SERCA mediated uptake ($v_{max}\_SERCA = 1.108$), and InsP$_3$ receptor (991.6666).

For both models, the time length is 20 seconds. The time step for the compartmental model is 0.1, and we keep every point. For the spatial models, use a time step of 0.001 and keep every 100th time point.

Running a simulation of a spatial model requires specifying a mesh size for the geometry. For the 2D geometry, the mesh assigns a grid of x,y coordinates to the existing geometry. The equations defining the behavior of the reactions are then solved within each cell of the mesh. The mesh size sets the size of the cells. A mesh of 0.2 μm has a finer resolution than one of 0.5 μm. The resolution obtained from optical studies is 0.2 μm. This is a minimum resolution that can be currently verified experimentally. The smaller mesh size also has a greater number of cells. The smaller the mesh size, the higher the resolution, the greater the number of cells, and the larger the number of computations required to solve the entire model. The results of a simulation can be sensitive to mesh size, similar to ODE approximations being sensitive to step size. The sensitivity of the model to mesh size can be tested by examining the results obtained using two different mesh sizes. In the Fink model and here, the 2D simulations were calculated within a domain of 7.7 × 103 and mesh size of 1.2 μm. Simulation results are named and stored for one's model and listed in the simulation view.

Detail 7.9

Setting a Mesh

When developing a model, start with a coarse grid size (60 × 40), and once the model is complete, increase to a finer resolution that may be more computationally intense (185 × 55). This can save computer run time and allows one to assess the sensitivity of the results to the mesh size.

7.4.4 Run and View Results

The previous pages stepped through creating one "Physiology" model and two applications to create two distinct complete models: "Compartmental" and "Uniform." In the following sections we will discuss the simulation results of these two models in relation to the experiments and simulations performed by Fink et al. (2000). The third simulation discussed in this section is the "Non-uniform" model. We do not create this application. Instead, we will use the existing Fink *et al.* BioModel to examine the results obtained when non-uniform distributions are taken into account. This third model is stored in the BioModel Shared Folder, CellBioGuide.

7.4.4.1 Compartmental Model

Expected Results The compartmental model serves as our baseline simulation. This simulation only takes into account the biochemical reactions of the model. It is possible and often the case that when a model is initially constructed, it fails to work. The question for the modeler is why? As a first order of business, check the parameters, initial conditions, and rate equations. It is fairly easy to create data entry errors with a physiologic model of 8 reactions and more than 30 numerical values. The next check is on the behavior of the rate equations. This is done two ways. First, we run a steady-state simulation to determine if the set of equations can reach a steady state. Given the biological system reaches a dynamic equilibrium, we expect to see a steady-state solution from the simulation. For this model, steady state is assumed to be reached when no stimulus is provided ($J_s = 0$). Next we can ask, do the changes in rates of reactions behave as expected? Although it is not intuitive how a variable changes over time in the context of multiple reactions, it is possible to have reasonable expectation for a single variable in relationship to a single reaction. Examining the relationship of a reaction rate to a variable can help the modeler determine whether it is behaving properly.

Some of the expected behaviors for the reactions in our model include the following:

Initial increase in InsP$_3$ in the cytosol and subsequent decrease over time.

Decreased rate of InsP$_3$ production and increased rate in InsP$_3$ degradation rate.

Increase flux through SERCA pump as Ca_Cytosol levels increase.

Simulation Results   To examine the behaviors identified above, we examine individual variables against time or another variable within the plot view of the data viewer (Fig. 7.14). For example, we can look at the change in concentration and rate of production or degradation of InsP$_3$ over time. We can also examine the change in SERCA flux in relation to changes in concentration of Ca_Cytosol.

Detail 7.10

Simulation Results: View Data, Compartmental

The results of completed simulations are viewed by clicking the "Results" tab. The relationship between any two variables can be plotted in the graph view. All data can be viewed as a table by using icons to toggle between graph or table views. Results can also be exported to a zip file.

In the experimental system, *in vivo* calcium levels increase 20-fold from 0.05 μM to 1 μM in the presence of 500 nM bradykinin (BDK) (Fink *et al.*, 1999b). In Fink *et al.* (2000), free cytosolic calcium reached a peak of 1.26 μM at 3 seconds in a simulation with fixed time steps, a slightly greater than 20-fold increase. In our compartmental simulation, we see calcium increase from 0.05 to 4.1 μM within a second (Fig. 7.15). This is nearly an 80-fold increase in cytosolic calcium levels.

Detail 7.11

Simulation Results: View Data, Spatial

The geometry of the simulation is shown in the data viewer. Each image is a single time point with the concentration of the selected variable at each *x,y* coordinate displayed. A slider is provided to look at the concentration distribution at different time points

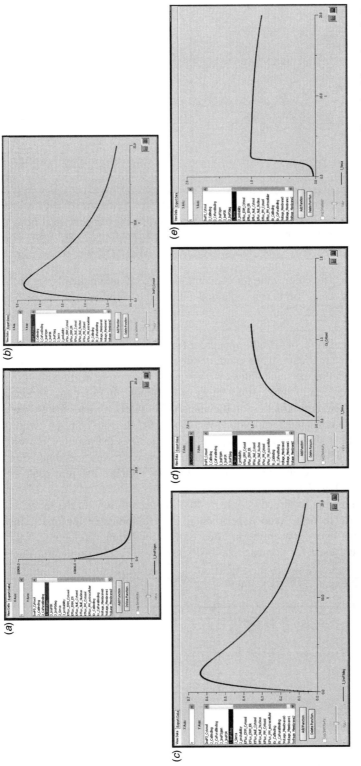

Figure 7.14 (a) InsP$_3$ generation over time. (b) InsP$_3$ concentration in the cytoplasm over time. (c) InsP$_3$ degradation over time. Flux of calcium through the SERCA pump in relation to time (d) and cytoplasmic calcium concentration (e).

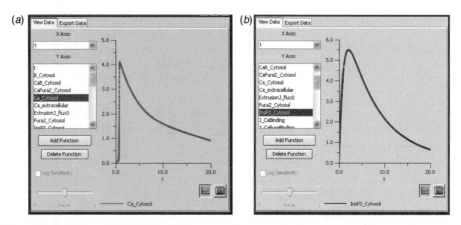

Figure 7.15   Calcium (a) and InsP₃ (b) concentrations over time in compartmental model simulation. Calcium (a) and InsP₃ (b) concentrations over time in compartmental model simulation with corrected rate constants for V_{max}, J_s, and J_{max}.

for a variable. The color scale for concentrations is set automatically for each variable based on the range of values obtained during the simulation and can be changed manually.

From this simulation, our results appear far from both the initial simulation designed by Fink and experimental data. If we examine the simulation "Parameters," we can see that the rate constants are the same values we assigned. They were not modified to take into account the volume ratio conversion factors of the model. In order for the kinetic model to describe the enzyme kinetics of the membrane reactions and fluxes, we must re-calculate the rate constants (Appendix). We can "Edit" the rate constants within the parameter view of the simulation control panel. This modifies the parameters without changing the BioModel or losing previous results. When we re-run the simulation with the newly calculated rate constants (see Appendix), we obtain a calcium peak of 1.19 μM at 2.86 seconds (Fig. 7.16).

It was not possible to directly compare InsP₃ concentrations in the model to those in the cell because there was no experimental *in vivo* indicator for InsP₃. Instead, the average cellular InsP₃ concentration after BDK stimulation was determined, 2 μM in whole cells, and used for comparison. InsP₃ levels in Fink's simulation rise from the

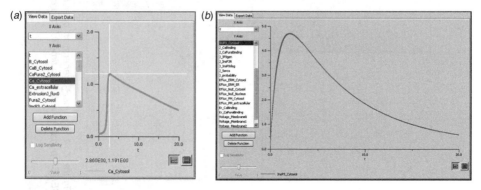

Figure 7.16   Calcium concentrations (a) and InsP₃ concentrations (b) versus time for compartmental model once correction factors are taken into account.

TABLE 7.12 Comparison of Calcium and InsP₃ Results from Compartmental Simulations

Species	Source	Peak (μM)	Time After Activation (s)
Calcium (init = 0.05 μM)			
	In vivo[a]	1.0–1.2	2–4
	Original	4.1	0.9
	Corrected	1.19	2.86
InsP₃ (init = 0.16 μM)			
	In vivo[a]	2.1	10
	Original	5.5	2
	Corrected	4.69	2
	Corrected	1.84	10

[a]*In vivo* experimental data reported in Fink *et al.* (1998, 2000).

initial concentration of 0.16 to 4.69 μM by 2.02 seconds (Table 7.12). When the unconverted parameter value is used, the model reaches a peak concentration of 5.5 μM in InsP₃ within 2 seconds, whereas in the model with the converted rate constants, InsP₃ concentrations increase from 0.16 μM to the peak concentration of 4.69 μM within 2 seconds. In both cases, the peak concentration is greater than the experimental whole cell measurement of 2.1 μM that was taken at 10 seconds. The 10-second value of the corrected simulations is lower than what was found *in vivo*.

Detail 7.12

Point Tool: Species Over Time

We use the point tool to look at data within a single location across time. This generates a time plot. The data associated with the selected *x,y* coordinate is sent to a graphical window where it can be viewed as a plot of values over time or as a table of values.

We can deduce from the initial compartmental simulations that the rate equations properly describe the behavior of their respective reactions, and InsP₃ dynamics are modeled sufficiently to match peak values in experimental results. Calcium dynamics showed discrepancies between simulation and experimental values that were addressed by taking into account the conversion factor for surface-to-volume and volume fractions. Thus, the model re-creates at a macroscopic and quantitative level changes in variable concentrations similar to what is seen *in vivo*.

7.4.4.2 Uniform Model

Because the model above is a "compartmental" model that does not explicitly model the cellular architecture, we do not have information about where in the cell these peaks are reached. A more detailed examination of the relationship between InsP₃ and calcium dynamics would include a comparison between simulation and experiments on the delay between peak calcium concentrations in neurite and soma in addition to the peak concentration values. Presumably, when this same model containing the machinery and dynamics of the system are executed within the cellular geometry, we will see propagation of the calcium wave similar to what we see *in vivo*: initiation at the neurite within 2 to 3 seconds and propagation to the soma and distal terminal. To test this, we examine the same physiologic model in the geometry of the neuroblastoma. Although the physiologic model is the same, we will no longer be performing a compartmental model. The cellular structures defined in the physiologic model are now mapped to geometric structures in the neuroblastoma image, and diffusion is enabled for the molecules that have diffusion coefficients (i.e., InsP$_3$, calcium, buffer, and Fura-2). We are now solving a set of partial differential equations. The default assumption in our spatial simulation is that molecules and structures are distributed uniformly throughout the 2D space.

Data Viewing   The "Data Viewing" tool for spatial simulation displays the geometry to which the model has been applied. This provides a visual representation of the variable values in the context of the 2D geometry. Data is available for every time point collected during the course of the simulation. In addition to the default display of 2D space at a specific time point, it is possible to export the simulation data into multiple formats including movie formats. The movie files are helpful for obtaining a qualitative understanding of changes in a variable over time or in variables in relation to one another. Quantitative values are more easily extracted from the default view or an exported ASCII file.

Simulation Results   We can now compare simulation and experimental results for calcium and InsP$_3$ dynamics within the soma and neurite (Fig. 7.17).

Experimentally, neurite spikes occur within 2.5 to 3 seconds after stimulation with BDK. The lag between neurite and soma peaks is very short within a second (Fink *et al.*, 2000). In our simulated compartmental model, calcium levels initiated and peaked first in the neurite followed by the soma (Table 7.13). The temporal lag

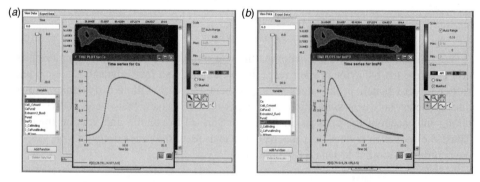

Figure 7.17 *Two-dimensional simulation with uniform distribution of organelles and molecules assumed. Results are presented in the data viewer with a scroll bar for scanning through temporal slices in the upper left, a list of variables including fluxes and reactions, and the cell geometry at the top center of the window. Data collection points were specified within the soma and neurite of the cell with the point tool. Graphs were produced by choosing variables to monitor, calcium (a) and InsP3 (b), and selecting the "Show Time Plots."*

TABLE 7.13 Calcium and InsP₃ Results from Uniform Spatial Simulation

Location	Species	Peak (μM)	Time After Activation (s)
Neurite *in vivo*	Calcium[a]	1.0	2–4
Neurite	InsP₃	6.3	1.7
Neurite	Calcium	1.01	3.7
Soma *in vivo*	Calcium[a]	1.0	2–4
Soma	InsP₃	2.47	2.4
Soma	Calcium	0.62	9.2

[a]*In vivo* experimental data reported in Fink *et al.* (1998, 2000).

between the calcium peaks in the neurite and soma was >5 seconds, more than a 10-fold increase compared with experimental values. As well, the peak calcium in the soma was less than what has been seen *in vivo*, which peaked at 0.62 μM versus the 1 μM *in vivo*. The temporal and spatial increases in InsP₃ preceded calcium increases in both the neurite and soma. This is consistent with the hypothesis that InsP₃ dynamics determine calcium patterns.

Nonetheless, the model is still not consistent with experimental results. The maximum calcium concentrations in the soma and neurite were lower than seen experimentally, and although increases in calcium move from neurite to soma, there was a significant change in the time it takes for the wave to move from neurite to soma. Why? What may be responsible for the difference between the experimental results and the simulation?

7.4.4.3 *Non-uniform*

Because the simulation results did not reflect what we see through experimentation, and we believe that the kinetic descriptions are accurate, we can conclude that the previous implementation of the model is insufficient to explain the spatial, temporal patterns seen in the cell. Fink *et al.* (2000) next hypothesized that a non-uniform distribution of the molecular components was required for proper

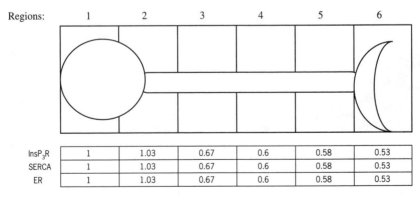

Figure 7.18 *Diagram illustrating the method of Fink et al. for obtaining relative fluorescent intensities of ER, SERCA, and InsP₃R over the body of the cell. Fluorescence intensity of ER in the soma was taken to be a value of 1. All other values were obtained relative to the soma. Values for each region are listed in this figure.*

calcium dynamics. To test this, Fink *et al.* used data collected on the distribution of organelles and species. They evaluated the density of InsP₃ receptors and SERCA pumps in six regions that spanned the length of the cells (Fig. 7.18). Density was calculated by measuring relative fluorescent intensities (r.i.), and the average density value per region was used as a factor in model simulations (Fink *et al.*, 1999b).

The ER density was found to vary across the length of the cell in a consistent fashion. Density based on relative fluorescence ranged nonlinearly from 0.95 r.i. in the distal neurite to 2.45 r.i. in the midneurite region. The molecular species, InsP₃, and SERCA pump were found to have the same relative fluorescent intensities as the ER within which they are located. As such, Fink *et al.* added a single density factor to the ER membrane (ER_density) and reactions. The third simulation we examine includes the geometry of the cell, the distributions of the receptor, ER, pumps, and channels within the geometry, and the biochemical reactions (Fig. 7.19).

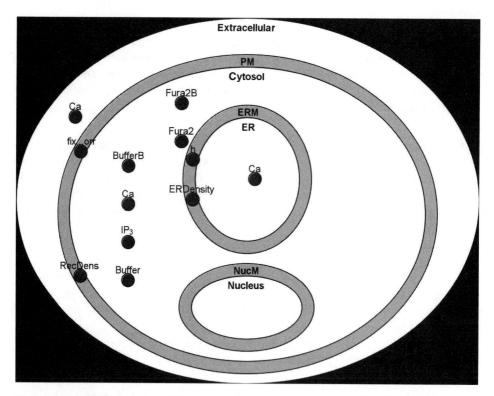

Figure 7.19 *The image of the "Physiology" model corresponding to the MathModel created by Fink et al. (2000) was obtained by using the "Save as image" feature available within Virtual Cell. This model was used to simulate non-uniform distributions within a spatial model. The species notation has been retained in the image (Fura2B is CaFura2; BufferB: CaB; IP3: InsP₃). Note the inclusion of two species: ERDensity and 2Dcorrection. The Fink model includes reactions for leakage of calcium across membranes. This is visible in the Application Pallet and has been disabled in the Reaction Mapping view in order to make comparisons across the three models in this chapter.*

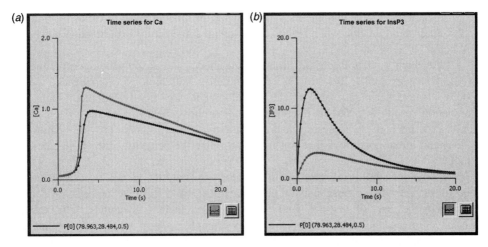

Figure 7.20 *Calcium (a) and InsP₃ (b) concentrations in soma and neurite from spatial simulation with nonuniformly distributed endoplasmic reticula and its associated membrane proteins. Graph data was collected from the same location (x,y coordinates) within soma and neurite as the simulation with uniform distributions.*

Spatial Models: Regions To assign region-specific values for the density factor, sub-regions were defined relative to the cell geometry. Conceptually, the regions are specified by coordinates based on the cell geometry, size, and point of origin. Fink *et al.* (2000) used the coordinates to include a series of conditional statements in the IC expression field that set the ER density to experimentally determine density values.

Simulation Results We can select the same coordinates to examine calcium and InsP₃ concentrations in the Non-uniform model as in the Uniform application because they use the same image file for their geometry. When we plot the simulation results of the Non-Uniform Fink's model, we can see that the concentrations of calcium reach amplitudes of 1.0 to 1.2 in the neurite and soma, respectively (Fig. 7.20). Significantly closer to the experimentally measured values. The temporal lag in calcium peaks between the neurite and soma peaks are returned to 0.2–0.4 seconds versus the 4-second lag seen in the previous simulation where distributions were not taken into account.

7.5 CONCLUSION

A computational model is composed of a set of hypotheses. The inclusion of species and the choice of kinetic types are each hypotheses about sufficient and required factors and biochemical reactions. The factors and rate equations ideally are quantitatively determined through experimental studies and used to reconstruct the system as a computational model. As a set of hypotheses concerning what is sufficient to create the experimental results, the hypotheses are false when the model fails to re-create the biological observations and true when results are re-created and future experimental behaviors predicted.

Of the three models discussed in-depth, the model and accompanying experimental studies most exemplify the process of quantitative cell biology (Slepchencko *et al.*, 2003). The experiments and model development are tightly coupled such that simulation results that fail to re-create the quantitative behavior of the system drive new experimental hypotheses, and quantitative experiments provide parameter values and initial conditions that constrain and refine the behavior of the computational model.

The series of studies examined in this chapter illustrate the successful use of modeling to develop better mechanistic understandings of a cellular process. We have described the use of Virtual Cell in analyzing calcium dynamics. The models we have created in this chapter are fairly complex. They take into account the temporal and spatial dynamics of both calcium and $InsP_3$. Each model is the implementation of the mathematical representation of components and the interactions believed to be required and sufficient for calcium dynamics in response to bradykinin stimulation.

The initial model failed to reproduce experimental results; the second model came close but had significant discrepancies. The third model, which was refined further with measured values for molecular distributions, reproduced the experimental results. By linking biochemical reactions to models of cellular morphology, Fink *et al.* were able to elucidate additional factors that play a significant role in the determination of calcium dynamics within neuroblastoma cells. The models have been further used to accurately predict the calcium dynamics in experiments that added a second buffer, CG-1 dextran, or globally increased $InsP_3$ concentrations in uncaging experiments (Fink *et al.*, 1999a).

Cells have a diversity of cellular shape and cytoplasmic architecture. Distribution and localization of organelles and molecular species contribute to the dynamic behaviors seen in response to external signaling cues. The degree to which these spatial characteristics participate in the quantitative and qualitative behavior is a question that can be addressed through the combined use of quantitative experimental and computational methods. The flexible design of Virtual Cell enables the construction and investigation of multiple hypotheses. That a single physiologic model can be applied to multiple geometries supports the evaluation of hypotheses regarding the relationship between spatial properties of the cell and biochemical reactions.

The benefits of creating such an extensive model in Virtual Cell are (1) multiple unique initial conditions can be examined by making new applications, (2) the role of spatial designs can be explored by applying distinct geometries, (3) the model can be shared with collaborators via the model database, and (4) the assembled models can be used for generating testable predictions and hypotheses about the biology.

7.6 APPENDIX: MODEL EQUATIONS AND PARAMETERS

Cellular Process	Syntax (Math, Virtual Cell)	Parameters[a] (reference, values)
	Differential equation for rate of change of calcium	
	$\dfrac{d[Ca_{cyt}]}{dt} = -(\text{rate of extrusion} + \text{rate of ER uptake} + \text{rate of buffering}) + \text{rate of release}$	
Extrusion: Cytosol to Extracellular space	$J_{max}[Ca]$, when $[Ca] > \text{threshold}$	Herrington et al., 1996 estimate
	$J_{max}*([Ca\_Cytosol] - \text{Threshold})*([Ca\_Cytosol] > \text{Threshold})$	$J_{max} = 8~\mu M\,s^{-1}$ Threshold $= 0.2~\mu M$
SERCA: ER uptake of Ca^{2+}	$J_{max}\dfrac{[Ca_{cyt}]^2}{[Ca_{cyt}]^2 + K_d^2}$	Fink et al., 2000
	$J_{max}*(Ca\_Cytosol \wedge 2.0)/((Ca\_Cytosol \wedge 2.0 + (K_d \wedge 2.0))$	$J_{max} = 3.75~\mu M\,s^{-1}$
Ca binding and release endogenous buffer	$k_{on}[Ca][B] - k_{off}[CaB]$	$k_r = K_{eq}*k_f$
	Mass Action $k_f[Ca\_Cytosol][B\_Cytosol] - k_r[CaB\_Cytosol]$	$K_{eq} = 10~\mu M$
		$k_f = 500$
Ca binding and release exogenous buffer (Fura 2)	$k_{on}[Ca][Fura2] - k_{off}[CaFura2]$	Fink et al., 2000
	Mass Action $k_f[Ca][Fura2] - k_r[CaFura2]$	$k_r = K_{eq}*k_f$
		$K_{eq} = 0.24~\mu M$
		$k_f = 500$
InsP$_3$ Receptor: ER release of Ca^{2+}	$J_{max}\left[\left(\dfrac{[InsP_3]}{[InsP_3] + K_{InsP_3}}\right)\left(\dfrac{[Ca]}{[Ca] + K_{act}}\right)h\right]^3\left(1 - \dfrac{[Ca]}{[Ca_{ER}]}\right)$	Fink et al., 2000
		Fink et al., 2000
		Fink et al., 2000
	$-J_{max}*(1.0 - (Ca\_Cytosol/Ca\_ER))*((InsP_3\_Cytosol/$ $(InsP_3\_Cytosol + K_{InsP_3}))*(Ca\_Cytosol/$ $(Ca\_Cytosol + K_{act}))*h\_ERM)\wedge 3.0$	$J_{max} = 3500~\mu M\,s^{-1}$ $K_{InsP_3} = 0.8~\mu M$ $K_{act} = 0.3~\mu M$
Probability of InsP$_3$ receptor opening (h_ERM)	$k_{on}(K_{in} - ([Ca]_{cyt} + k_{in})*h)$	Fink et al., 2000
		Fink et al., 2000
	$K_{on}*(K_{in} - ((Ca\_Cytosol + K_{in})*h\_ERM))$	$k_{on} = 2.7~\mu M\,s^{-1}$ $K_{in} = 0.2~\mu M$

(Continued)

Appendix: *Continued*

Cellular Process	Syntax (Math, Virtual Cell)	Parameters[a] (reference, values)
Differential equation for the rate of change of InsP$_3$ $\quad InsP_3 \dfrac{d[InsP_3]}{dt}$ = rate of production – rate of degradation		
InsP$_3$ generation	$602 * J_{InsP_3}\, e^{-kt}$ $602 * J_{InsP_3} * \exp(-k*t)$	$J_{InsP_3} = 20.86\ \mu M\ s^{-1}$ $k = 1.188\ s^{-1}$ t = time
InsP$_3$ degradation	$-k(InsP_{3cyt} - InsP_{3init})$ $(k*(InsP_{3\_}Cytosol - InsP_{3\_}initial))$	$k = 0.14\ s^{-1}$ $InsP_{3init} = 0.16\ \mu M$

[a]Rate constants for membrane bound enzimex are recalculated to take into account correction factors in spatial simulations.

Recalculated Rate Constants for Spatial Simulations

Mechanism	Original Rate Constant	Corrected Rate Constant Value
Extrusion pumps	$J_{max} = 8 \ \mu Ms^{-1}$	$6.8 \ \mu Ms^{-1}$
SERCA pump	$J_{max} = 3.75 \ \mu M \ s^{-1}$	$1.0625 \ \mu M \ s^{-1}$
InsP$_3$ receptor	$J_{max} = 3500 \ \mu M \ s^{-1}$	$991.6666 \ \mu M \ s^{-1}$
InsP$_3$ generation	$J_{InsP_3} = 20.86 \ \mu M \ s^{-1}$	$17.7309 \ \mu M \ s^{-1}$

BIBLIOGRAPHY

Allbritton NL, Meyer T, Stryer L (1992). Range of messenger action of calcium ion and inositol 1,4,5-trisphosphate. *Science* 258:1812–1815.

Atri A, Amundson J, Clapham D, *et al.* (1993). A single-pool model for intracellular calclium oscillations and waves in the *Xenopus laevis* oocyte. *Biophysical Journal* 65(4):1727–1739.

Berridge MJ (1993). Inositol trisphosphate and calcium signalling. *Nature* 361: 315–325.

Berridge MJ, Irvine RF (1984). Inositol trisphosphate, a novel second messenger in cellular signal transduction. *Nature* 312:315–321.

Bezprozvanny I, Watras J, Ehrlich BE (1991). Bell-shaped calcium-response curves of Ins(1,4,5)P$_3$- and calcium-gated channels from endoplasmic reticulum of cerebellum. *Nature* 351:751–754.

Coggan JS, Thompson SH (1995). Intracellular calcium signals in response to bradykinin in individual neuroblastoma cells. *American Journal of Physiology* 269:C841–848.

Coggan JS, Thompson SH (1997). Cholinergic modulation of the Ca^{2+} response to bradykinin in neuroblastoma cells. *American Journal of Physiology* 273:C612–617.

DeLisle S, Welsh MJ (1992). Inositol trisphosphate is required for the propagation of Ca^{2+} waves in *Xenopus* oocytes. *Journal of Biological Chemistry* 267:7963–7966.

De Young GW, Keizer J (1992). A single-pool inositol 1,4,5-triphosphate-receptor-based model for agonist-stimulated oscillations in Ca^{2+} concentration. *Proceedings of the National Academy of Sciences USA* 89:9895–9899.

Farooqui AA, Anderson DK, Flynn C, *et al.* (1990). Stimulation of mono- and diacylglycerol lipase activities by bradykinin in neural cultures. *Biochemical and Biophysical Research Communications* 166(2):1001–1009.

Fink C, Morgan F, Loew LM (1998). Intracellular fluorescent probe concentrations by confocal microscopy. *Biophysical Journal* 75:1648–1658.

Fink CC, Slepchenko B, Loew LM (1999a). Determination of time-dependent inositol-1,4,5-triphosphate concentrations during calcium release in a smooth muscle cell. *Biophysical Journal* 77(1):617–628.

Fink CC, Slepchenko B, Moraru II, *et al.* (1999b). Morphological control of inositol-1,4,5-triphosphate-dependent signals. *Journal of Cell Biology* 147(5):929–936.

Fink CC, Slepchenko B, Moraru II, *et al.* (2000). An image-based model of calcium waves in differentiated neuroblastoma cells. *Biophysical Journal* 79:163–183.

Gill DL, Grollman EF, Kohn LD (1981). Calcium transport mechanisms in membrane vesicles from guinea pig brain synaptosomes. *Journal of Biological Chemistry* 256(1):184–192.

Gill D, Chueh SH (1985). An intracellular (ATP + Mg^{2+})-dependent calcium pump within the N1E-115 neuronal cell line. *Journal of Biological Chemistry* 260:9289–9297.

Glanville NT, Byers DM, Cook HW, *et al.* (1989). Differences in the metabolism of inositol and phosphoinositides by cultured cells of neuronal and glial origin. *Biochimica et Biophysica Acta* 1004(2):169–179.

Herrington J, Park JB, Babcock DF, *et al.* (1996). Dominant role of mitochondria in clearance of large Ca^{2+} loads from rat adrenal chromaffin cells. *Neuron* 16:219–228.

Higashida H, Brown DA (1987). Bradykinin inhibits potassium (M) currents in N1E-115 neuroblastoma cells. Responses resemble those in NH108–15 neuroblastoma x glioma hybrid cells. *FEBS Letters* 220(2):302–306.

Hirose K, Kadowaki S, Tanabe M, *et al.* (1999). Spatiotemporal dynamics of inositol 1,4,5-trisphosphate that underlies complex Ca^{2+} mobilization patterns. *Science* 284:1527–1530.

Iredale P, Martin K, Hill S, *et al.* (1992). Agonist-induced changes in [Ca^{2+}]$_i$ in N1E-115 cells: differential effects of bradykinin and carbachol. *European Journal of Pharmacology* 226(2): 163–168.

Jaffe LF (1983). Sources of calcium in egg activation: a review and hypothesis. *Developmental Biology* 99:265–276.

Jaffe LF (1991). The path of calcium in cytosolic calcium oscillations: a unifying hypothesis. *Proceedings of the National Academy of Sciences USA* 88:9883–9887.

Jafri MS (1995). A theoretical study of cytosolic calcium waves in *Xenopus* oocytes. *Journal of Theoretical Biology* 172(3):209–216.

Kasai H, Li YX, Miyashita Y (1993). Subcellular distribution of Ca^{2+} release channels underlying Ca^{2+} waves and oscillations in exocrine pancreas. *Cell* 74(4):669–677.

Keizer J, De Young GW (1992). Two roles of Ca^{2+} in agonist stimulated Ca^{2+} oscillations. *Biophysical Journal* 61(3):649–660.

Keizer J, Levine L (1996). Ryanodine receptor adaptation and Ca^{2+}(-)induced Ca^{2+} release-dependent Ca^{2+} oscillations. *Biophysical Journal* 71(6):3477–3487.

Klingauf J, Neher E (1997). Modeling buffered Ca^{2+} diffusion near the membrane: implications for secretion in neuroendocrine cells. *Biophysical Journal* 72:674–690.

Kupferman R, Mitra PP, Hohenberg PC, *et al.* (1997). Analytical calculation of intracellular calcium wave characteristics. *Biophysical Journal* 72(6):2430–2444.

Lechleiter JD, Girard S, Peralta E, *et al.* (1991). Spiral calcium wave propagation and annihilation in *Xenopus laevis* oocytes. *Science* 252:123–126.

Li YX, Rinzel J (1994). Equations for InsP$_3$ receptor-mediated [Ca^{2+}]$_i$ oscillations derived from a detailed kinetic model: a Hodgkin-Huxley-like formalism. *Journal of Theoretical Biology* 166:461–473.

Li YX, Keizer J, Stojilkovic SS, *et al.* (1995). Ca^{2+} excitability of the ER membrane: an explanation for IP3-induced Ca^{2+} oscillations. *American Journal of Physiology* 269: C1079–1092.

Lytton J, Westlin M, Burk SE, *et al.* (1992). Functional comparisons between isoforms of the sarcoplasmic or endoplasmic reticulum family of calcium pumps. *Journal of Biological Chemistry* 267:14483–14489.

Mathes C, Thompson SH (1994). Calcium current activated by muscarinic receptors and thapsigargin in neuronal cells. *Journal of General Physiology* 104:107–121.

Meldolesi J, Pozzan T (1998). The endoplasmic reticulum Ca^{2+} store: a view from the lumen. *Trends in Biochemical Sciences* 23:10–14.

Meyer T, Holowka D, Stryer L (1988). Highly cooperative opening of calcium channels by inositol 1,4,5-trisphosphate. *Science* 240:653–656.

Milner RE, Famulski KS, Michalak M (1992). Calcium binding proteins in the sarcoplasmic/endoplasmic reticulum of muscle and nonmuscle cells. *Molecular and Cellular Biochemistry* 112(1):1–13.

Miyawaki A, Llophis J, Heim R, *et al.* (1997). Fluorescent indicators for Ca^{2+} based on green fluorescent proteins and calmodulin. *Nature* 388:882–887.

Neher E, Augustine GJ (1992). Calcium gradients and buffers in bovine chromaffin cells. *Journal of Physiology* 450:273–301.

Nuccitelli R, Yim DL, Smart T (1993). The sperm-induced Ca^{2+} wave following fertilization of the *Xenopus* egg requires the production of Ins(1,4,5)P3. *Developmental Biology* 158:200–212.

O'Sullivan AJ, Cheek TR, Moreton RB, *et al.* (1989). Localization and heterogeneity of agonist-induced changes in cytosolic calcium concentration in single bovine adrenal chromaffin cells from video imaging of fura-2. *EMBO Journal* 8(2):401–411.

Parker I, Ivorra I (1990). Localized all-or-none calcium liberation by inositol trisphosphate. *Science* 250(4983):977–979.

Reiser G, Cesar M, Binmoller FG (1992). Bradykinin and muscarine induce Ca^{2+} dependent oscillations of membrane potential in rat glioma cells indicating a rhythmic Ca^{2+} release from internal stores: thapsigargin and 2,5-D(tert-butyl)-1,4, benzohydroquinone deplete IPs-sensitive Ca^{2+} stores in glioma and in neuroblastoma-glioma hybrid cells. *Experimental Cell Research* 202:440–449.

Richelson E (1979). Tricyclic antidepressants and histamine H1 receptors. *Mayo Clinic Proceedings* 54(10):669–674.

Rooney TA, Sass EJ, Thomas AP (1990). Agonist-induced cytosolic calcium oscillations originate from a specific locus in single hepatocytes. *Journal of Biological Chemistry* 265(18):10792–10796.

Roth BJ, Yagodin SV, Holtzclaw L, *et al.* (1995). A mathematical model of agonist-induced propagation of calcium waves in astrocytes. *Cell Calcium* 17(1):53–64.

Slepchenko BM, Schaff JC, Macara I, *et al.* (2003). Quantitative cell biology with the Virtual Cell. *Trends in Cell Biology* 13(11):570–576.

Sneyd J, Keizer J, Sanderson MJ (1995). Mechanisms of calcium oscillations and waves: a quantitative analysis. *FASEB Journal* 9:1463–1472.

Snider R, Richelson E (1984). Bradykinin receptor-mediated cyclic GMP formation in a nerve cell population (murine neuroblastoma clone N1E-115). *Journal of Neurochemistry* 43(6):1749–1754.

Surichamorn W, Forray C, el-Fakahany EE (1990). Rough of intracellular Ca^{2+} mobilization in muscarinic and histamine receptor-mediated activation of guanylate cyclase in N1E-15 neuroblastoma cells: assessment of the arachidonic acid release hypothesis. *Molecular Pharmacology* 37(6):860–869.

Tang Y, Othmer HG (1994). A model of calcium dynamics in cardiac myocytes based on the kinetics of ryanodine-sensitive calcium channels. *Biophysical Journal* 67(6):2223–2235.

Tse A, Tse FW, Hille B (1994). Calcium homeostasis in identified rat gonadotrophs. *Journal of Physiology* 477(3):511–525.

Ueda T, Chueh SH, Noel MW, *et al.* (1986). Influence of inositol 1,4,5-trisphosphate and guanine nucleotides on intracellular calcium release within the N1E-115 neuronal cell line. *Journal of Biological Chemistry* 261(7):3184–3192.

Wagner J, Keizer J (1994). Effects of rapid buffers on Ca^{2+} diffusion and Ca^{2+} oscillations. *Biophysical Journal* 67:447–456.

Wang SSH, Thompson SH (1995). Local positive feedback by calcium in the propagation of intracellular calcium waves. *Biophysical Journal* 69:1683–1697.

Wang SSH, Alousi AA, Thompson SH (1995). The lifetime of inositol 1,4,5-trisphosphate in single cells. *Journal of General Physiology* 105:149–171.

Xu T, Naraghi M, Kang H, *et al.* (1997). Kinetic studies of Ca^{2+} binding and Ca^{2+} clearance in the cytosol of adrenal chromaffin cells. *Biophysical Journal* 73:532–545.

Ziche M, Zawieja D, Hester RK, *et al.* (1993). Calcium entry, mobilization, and extrusion in postcapillary venular endothelium exposed to bradykinin. *American Journal of Physiology* 265:H569–H580.

Chapter *8*

Advanced Computing

8.1 ADVANCED COMPUTING CONCEPTS AND RESOURCES

The preceding chapters focused on two technical and methodological areas in computational biology. In the first chapters, we discussed the use of sequence and protein family databases to find similarities between molecular sequences—DNA, protein, mRNA. In subsequent chapters, we discussed mathematical models of cellular processes using simulation tools that create ordinary differential equations that are solved numerically. We also briefly introduced spatial models. The technologies discussed in these chapters run directly on your desktop or laptop using a single processor (Stella, Gepasi) or are accessed remotely via a Web site (NCBI, PFAM, Virtual Cell).

In this chapter, we extend these conversations in the following ways. We highlight strategies for database and workflow management for bioinformatic-related research that take advantage of distributed computing resources. We expand our discussion of spatially realistic models to the introduction of MCell, which has combined modeling high-resolution graphics with stochastic mathematical models. We introduce software languages designed to make numerical models accessible across simulation platforms and software frameworks that enable large and multiscale simulations. We discuss community efforts to make models broadly accessible for use in simulation tools through the development of Extensible Markup Languages (XMLs). Simulation frameworks are discussed for their ability to integrate models across scales (time and size) and type (continuous and discrete), as well as increase computational functionality for analysis and visualization.

8.2 STANDARDS FOR MODEL DATA EXCHANGE

The models we created in Chapters 4–6 made use of three tools, two tailored for biochemical and biological models, the other a general simulator. Multiple tools have been developed for creating kinetic models of biochemical networks. Twelve kinetic simulators were recently reviewed by Alves *et al.* (2006). These are only a fraction of the total number of simulators available for kinetic modeling of biochemical networks. The tools were developed independently of one another and have different strengths. Generally, they are all model biochemical reactions. More specifically, some perform metabolic control analysis, stochastic simulations, optimization algorithms, or parameter fitting (Alves *et al.*, 2006).

Researchers develop models in their favorite simulator. This choice may reflect any number of preferences:

Available approximation methods (e.g., Euler, Runge-Kutta, LSODE)
User interface design (e.g., workflow, icon-based, command line)
Required computing platform (PC, Mac, Linux, Web-based)
Ability to write or modify ODEs directly (Virtual Cell, MathModel)

To share models developed in one simulator with a researcher or educator who uses a different simulation tool is not trivial. It requires a method for transferring data—variables, rate equations, and parameters—from one program to another without losing, misplacing, or changing data.

XMLs provide a method for describing, defining, and handling data (Bray *et al.*, 2000). They create a set of standard data types that enable users to take a single data model and run it across multiple platforms. This is true in Web pages, mathematics, and now increasingly in biochemical simulators. The ability to export data in a format that can be readily used by another program allows researchers to take advantage of the functionality of multiple simulators.

Systems Biology Markup Language (SBML) and CellML are two XML standards designed to add interoperability to tools used for modeling cellular events (Cuellar *et al.*, 2003; Hucka *et al.*, 2003). They transport models from one tool to another. For example, a researcher might develop a model of a signaling pathway using the biochemical simulator Gepasi (Fig. 8.1). The researcher prefers Gepasi's user interface and wants to use the parameter estimation tools in Gepasi to determine the robustness of the model. However, the researcher also has images of the cell shape in which the reactions occur and wishes to determine the affect of the cellular geometry on the results of the simulation. Virtual Cell is this researcher's choice for doing the spatial simulation. Rather than rewriting the model in each tool, the model can be exported from Gepasi in SBML format and subsequently imported to Virtual Cell where the researcher can use the geometry and spatial modeling tools for further investigation. The researcher transitions from a nonspatial to spatial environment without the model having to be completely rewritten, and the researcher is able to take advantage of the functionalities in both software environments.

SBML has focused on the minimum yet essential characteristics of models and modeling tools required to pass a model from one tool to another (Hucka *et al.*, 2003). It assumes that the sufficient components for modeling are identification of compounds

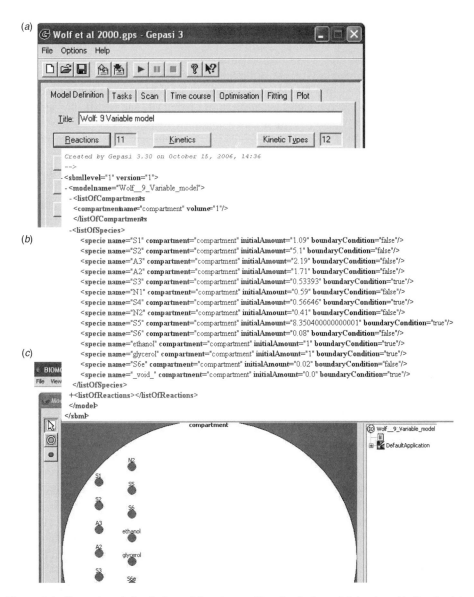

Figure 8.1 *Illustration of glycolysis model exchange. The glycolysis model developed in the simulator Gepasi (a) was exported to an SBML document (b). The compartmental model was then imported to Virtual Cell (c). Export and import commands are found in the File menus of the GUIs for Gepasi and Virtual Cell.*

(species); defining species (as reactants or products within a reaction); and assigning a theoretical space within which the reaction occurs (compartment).

CellML involves a larger set of metaconcepts for describing models (Cuellar *et al.*, 2003). In addition to Species, Reactions, and Compartments, CellML uses MathML within Reactions to describe the equation used for the kinetics. (This was in contrast to the use of text strings by SBML.) CellML also includes metadata tags to identify the pathway, functional role, and biological type (organ, fertilization, cell division, etc.). The metadata is a precursor for creating larger models of biological systems based on organizing biological functions into a hierarchical structure.

The scopes of the two XMLs are different. This leads to different yet overlapping definitions and relationships. The implementation of MathML for writing equations in SBML (Shapiro *et al.*, 2004) and the stabilization of CellML and SBML has led to the development of a conversion method that supports converting CellML models into SBML models (Schilstra *et al.*, 2006). This is consistent with the premise of XML efforts: exchangeable data.

Publicly available models in SBML can be imported into tools used in labs or classrooms as long as they can import the SBML format. CellML has a repository of models available at the CellML site along with tools capable of running the simulation (Jsim and mozCellML) (Miller *et al.*, 2000, 2001). Tools, like Virtual Cell, that maintain their own repository and export SBML or CellML create another resource for exploring and evaluating existing models. A list of model repositories can be found at systems-biology.org, a Web site sponsored by the Systems Biology Institute and Kitano Symbiotic Systems Project (http://www.systems-biology.org). Models are also now available via the European Bioinformatics Institute (EBI) Web site (http://www.ebi.ac.uk/biomodels; Le Nouvere *et al.*, 2006).

8.3   STRATEGIES FOR BIOINFORMATICS DATABASE AND RESOURCE/RESEARCH MANAGEMENT

Characterizing molecular properties of a newly isolated DNA sequence or full genomes requires running multiple independent algorithms often against several databases to detect and validate homology, structure, and function. Such a characterization might involve the following workflow: BLASTn/BLASTx to obtain putative protein sequence → CLUSTAL-W or HMMalign for a multiple alignment → PHYLIP for phylogenetic relationships → HMMer for detection of conserved domains → identification of known structure from PDB (Fig. 8.2). With the multitude of bioinformatics tools and databases, methods to manage databases and research resources are needed. A couple of strategies have been implemented. One involves the development of aggregated resources. Another involves the development of workflows that direct data from one tool to another.

Workflow management systems are a strategy that has been implemented for managing molecular research protocols that involve the use of multiple bioinformatics resources. The focus of workflow management systems is to automate the work process (Hollingsworth, 1998). Automation ensures that routines are performed in the same order with the same parameters from one experiment to another. This increases accurate comparison of results.

An example workflow could be that after isolating a new cDNA clone, you wish to BLAST against a nucleic acid database to obtain a set of homologous cDNAs. The retrieved cDNAs are then used in a multiple sequence alignment to determine what sequence segments are conserved between the cDNAs. The workflow management systems are designed to operationalize these steps by defining input and output, specifying which tools and parameter settings are to be used, and sending the output from one process to the next appropriate location (database, procedure, etc.). Workflow management

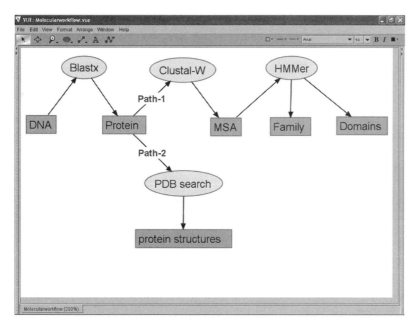

Figure 8.2 *Workflows for molecular sequence analysis. Two putative workflows starting with a DNA sequence are illustrated. Computational tools used to process sequence data are shown in circles, the type of data processed is shown in circles. The first workflow takes the DNA sequence, performs a BLASTx conversion to protein sequence, followed by creation of a MSA using CLUSTAL-W and subsequent searches for family or domain relationships. The second workflow uses the protein sequence from the second step to query the PDB records directly based on sequence similarity. Concept map was created using concept mapping tool VUE.*

systems for bioinformatics include methods for storing results in databases such that the data can be retrieved as input for additional analysis or queries; documenting parameters of methods (i.e., e-values, word size, gap penalties) and sequence of procedures (i.e., BLASTn followed by MSA). The same procedure can now be performed over and over again by the automated workflow.

8.3.1 Aggregating Resources and Data

The possibly more familiar approach is the aggregation of data from existing bioinformatics resources. Software programs and Web sites have been developed that bring together search algorithms and database. Suites of algorithms have been compiled into a set of tools accessible from a single site or CD, such as TM4 from The Institute for Genomic Research (TIGR) for microarray data processing and analysis. Databases have been integrated or federated on centralized and distributed computing platforms to provide an apparently seamless repository of multiple data types. UniProt, InterPro, Biology Workbench, and Entrez are just a few examples of such databases (Schuler *et al.*, 1996; Subramaniam, 1998; Bairoch *et al.*, 2005; Mulder *et al.*, 2005).

The Biology Workbench developed by Shankar Subramaniam has tools for searching, aligning, and viewing molecular sequence and structure data. The workbench provides a single user interface for access to databases and popular search algorithms (Fig. 8.3).

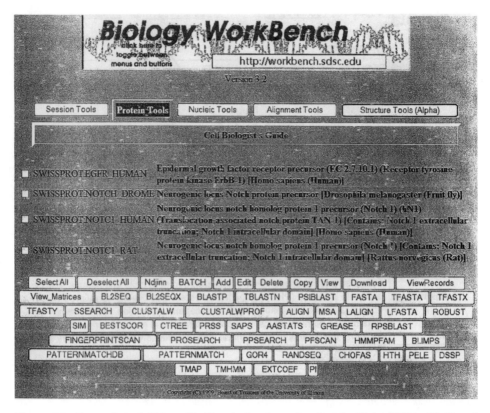

Figure 8.3 *Screenshot of Biology Workbench Session using Protein Tools. Biology Workbench organizes computational tools into four categories: Protein Tools, Nucleic Tools, Alignment Tools and Structure Tools. Searches and data are stored remotely within workbench sessions named by the user. The session shown is named "Cell Biologist's Guide." Four SwissProt records have been imported to the session for use with the protein tools. The integrated protein tools are visible at the bottom of the page (e.g., the BLAST protein algorithm and subroutines are visible in the second row of buttons).*

The site uses a series of perl scripts to run queries from the site to databases. The use of a single interface makes it possible for students and researchers to grapple with the biological data and search methods instead of navigation strategies and Web site design. Biology Workbench also provides a workspace for each user where results and queries can be stored with the research tool on the Biology Workbench server. Biology Workbench and its offshoot Biology Student Workbench have been used in professional development workshops for biology educators by the BioQUEST Curriculum Consortium and the Student Biology Workbench project of the University of Illinois at Urbana Champaign. The workshop directors and faculty participants have created multiple guides and curricular resources that can be used for introductory research exercises and undergraduate courses.

The Bioportal developed by the Renaissance Computing Institute (RENCI) at the University of North Carolina at Chapel Hill is similarly a tool that uses software programs to aggregate resources. The Bioportal provides access to more than 100 computational tools and more than 300 gigabytes (GB) of data in databases (Table 8.1). Installation of

TABLE 8.1   Computational Tools and Databases Available Through Bioportal

Database searching/sequence editing (32 tools)
Alignment and phylogeny (37 tools)
Pattern searching (12 tools)
DNA/RNA analysis (37 tools)
Protein analysis (23 tools)
NCBI database (95 GB)
GenBank (206 GB)
GenPept (3 GB)
PDB (6.3 GB)
PFAM (8.7 GB)
Prints (72 MB)
TransFac (36 MB)

the service on your own cluster requires a system administrator capable of configuring and managing cluster resources. The NC Bioportal team has developed a series of guides and downloads to assist research groups to establish their own Bioportal node capable of accessing Grid resources. The Grid, now also referred to as cyberinfrastructure, is the set of geographically distributed resources including hardware, data sets and people (Fig. 8.4). Access to computing resources is mediated by software protocols that authenticate user access to hardware and software, queue requested operations by software applications, and move data across computing platforms.

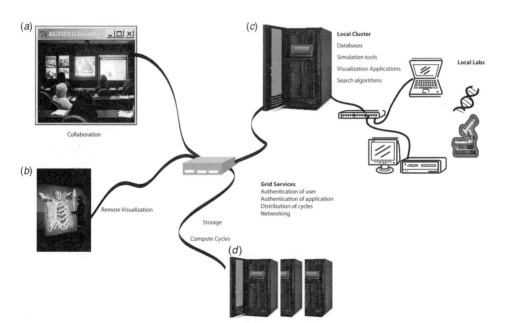

Figure 8.4 *The Grid is a set of geographically distributed resources accessed, navigated and managed over communication networks (black lines). The Grid enables people to collaborate via technologies such as the (a) Access Grid, to access (b) remote visualization and instrumentation tools, as well as compute and storage cycles from (c) local and (d) remote compute clusters. Grid Services provide researchers at local labs (your own lab or institutions) permission to and scheduling for use of multiple networks and compute resources.*

The Bioportal itself is a distributed system that uses geographically dispersed heterogeneous resources. The Bioportal services use grid computing protocols to enable the system to make use of remote TeraGrid—compute, storage, and algorithm—resources as needed (Fig. 8.5). To make use of the Bioportal, a lab creates its own bioportal compute cluster. Creating a local Bioportal cluster establishes a customizable dedicated resource for the research lab. Local instances allow the system administrator to manage accounts, computer permissions, and the compute time allowed for simulations or searches. This can decrease wait time and cues for computer access.

The ability to access resources beyond the locally dedicated cluster becomes important as the number and size of searches increase. For example, single sequence queries against multiple databases may take less than 5 seconds and use very little CPU time. However, when entire new genomes have been characterized (e.g., new bacterial strains), a genome-against-genome database query is compute-intensive and requires

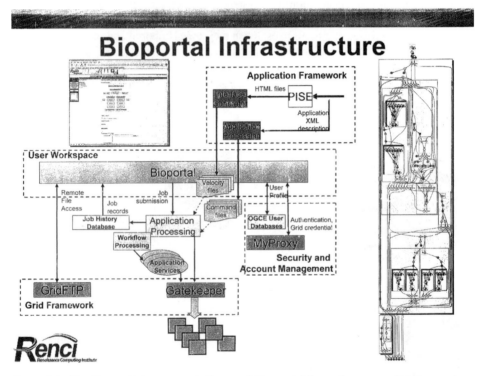

Figure 8.5 *The Bioportal infrastructure is illustrated. The top left image is a screenshot of a user interface to the Bioportal User Workspace. The User Workspace is the interface through which users interact with the Grid and web services that mediate interactions between machines across the internet. The Application Framework within Bioportal uses PISE (Letondal, 2001) to make bioinformatics tools and resources available to the user via Bioportal. The Grid Framework includes software that manages access to remote computers, execution of applications, monitoring of processes, and movement of data from one program or computer to another. Security and Account Management is managed via MyProxy which stores data about users including permissions for access to resources. An example workflow generated by Taverna is shown to the far right (Oinn et al., 2004). The workflow indicates the transfer of data from one program to another within the Bioportal. Schematic provided by North Carolina/TeraGrid Bioportal, Renaissance Computing Institute.*

additional CPU and memory resources. When researchers hit the limits of their compute resources, they must redesign their search strategy, forego the analysis, or modify their resources. Grid-enabled resources like Bioportal eliminate the need to make such choices by integrating high-performance protocols that allow access to resources as the need arises.

8.4 SIMULATION FRAMEWORKS FOR LARGE AND MULTISCALE MODELS

Tools capable of simulating multiscale models in real time use high-performance computing (HPC), collaborative tools, and methods from multiple fields (e.g., biology, computer science, applied mathematics). Advanced computing resources are needed in order to model at multiple resolutions, synchronize results, and visualize data across scales and within spatially realistic models. The interfaces for these tools (command line prompts, scripts or dataflow structures) are typically less intuitive to users who are not already familiar with programming and mathematical modeling. Developers are beginning to tailor interfaces to these advanced computing and visualization resources to domain areas based on the needs of the scientific user. The capabilities available to researchers via these tools are awesome and worth knowing about.

One scientific computing and visualization framework known as Problem-solving environment (PSE) that has been used with large biological models is DAFFIE (DAFFIE, 2006). Problem-solving environments are designed to bring together tools for 3D graphics, parallel computing, and numerical analysis to aid in the development of modeling environments and analysis of simulations. Problem-solving environments provide tools for transforming data values to optical properties, generating grids and meshes of spatial models, and optimizing algorithms. DAFFIE (Distributed Applications Framework for Immersive Environments) is the behind the scenes architecture for EarLab, a virtual laboratory developed for the integration of numerical models of auditory systems.

EarLab, a virtual environment for auditory research, is a framework for integration of multiple models within and across levels of abstraction. EarLab consists of modules that are used to perform virtual experiments. A virtual experiment is created by selecting modules for stimuli (sound files), one or more mathematical models of the biological structure (hair cells, basal membrane), analysis, and visualization (Fig. 8.6).

Conceptually, for biological models to talk to one another, an explicit relationship must exist between them. The models in EarLab are integrated hierarchically based on physiology and anatomy. The type of mathematical models can differ in their method and type of biological phenomenon. For example, the spiral structure of the cochlea contains along its length a basilar membrane (Fig. 8.7). Attached to the basilar membrane are inner hair cells that are innervated with 10–20 auditory nerves. The basilar membrane acts as a mechanical filter for sound. As sound hits the membrane, it is displaced at some frequency, which results in changes in the membrane potentials of the inner hair cells and subsequent action potentials in the auditory nerves. Each biological structure has a module in EarLab. The modules are their own mathematical models of each function.

Module parameter values are kept separate from the mathematical model. If the components of the ear are the same across mammalian species and what differs is the set of parameter values for those components, the models become species specific by entering the appropriate parameters at run time. By keeping parameters separate, the model can

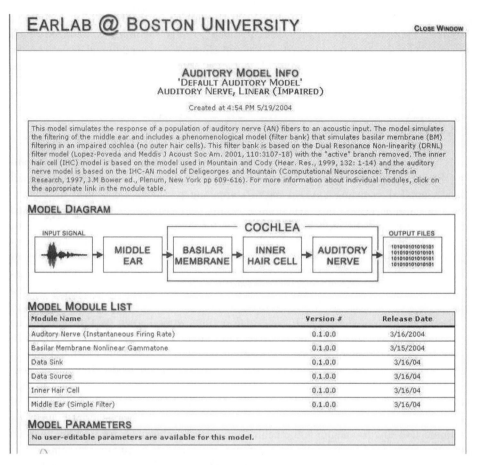

Figure 8.6   *EarLab modules for online simulation. The mathematical models in EarLab include detailed anatomic descriptions of cochlear tissues, explicit descriptions of the electrical properties of the tissues, and explicit modeling of the excitation of the nerve membrane of individual fibers (Girzon, 1987; Whiten, 2003). The default model within the online modeling suite is the AUDITORY NERVE. The model diagram highlights the modular components of a virtual experiment. The modules used in the default AUDITORY MODEL are listed in the Model Module List.*

be reconfigured for different mammalian species without having to rewrite the entire model. This achieves functionality analogous to the use of Applications in Virtual Cell (Chapter 7) where multiple parameter settings saved as independent initial conditions are applied to a single physiologic model of components and rate equations.

The EarLab architecture, DAFFIE, solves the set of modules in discrete time intervals. Each module computes for one time interval and then data is exchanged; results are sent from one module as outputs and received as new input for another. The data exchange at the end of each step synchronizes all modules in simulation time. The concept of frame synchronization is most easily illustrated with a sample sound input file (Fig. 8.8). For example, sound is a long wave form. The time it takes to say the word *Boston* is ~0.3 second. However, the time frames for module computations are 10 milliseconds. To synchronize the input with the time interval of other modules, the input signal is broken into smaller frames.

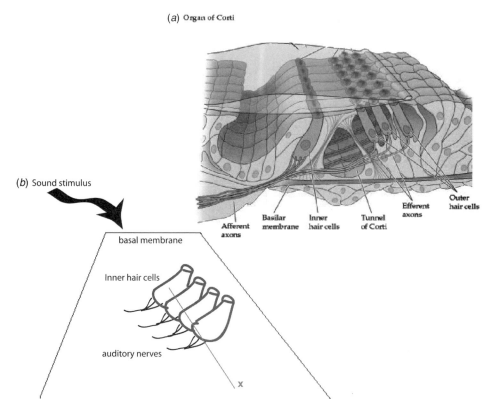

Figure 8.7 *(a) Illustration of Cochlea anatomic structure. (b) Simplified illustration of hearing developed by Greenwood (1961) that relates the characteristic frequency (CF) of any location along the length of the basilar membrane to the distance (x) of that location from the apex. $CF = A(10^{ax/L} - K)$: A is a constant that controls the high-frequency limit of the map (Hz); a is a constant that controls slope of the map; L is cochlear length in millimeters; K is a constant that controls low-frequency behavior.*

EarLab can be run as a desktop application or within DAFFIE. The DAFFIE architecture supports successful communication between modules and across computing platforms. The distributed implementation hosted at Boston University has a Web-based interface that allows one to configure and run models online via a Web browser (http://earlab.bu.edu/). Additional modules can be written to run in EarLab using the programming language of C or using MATLAB.

8.5 SPATIAL AND PROBABILISTIC MODELS

The modeling chapters in this book were written to progress from familiar simple models to more detailed models each using different biochemical simulators. We began with sets of ordinary differential equations that describe biochemical reactions occuring in a homogenous, uniform space and do not model spatial features. We next compared results between nonspatial and spatial models of biochemical reactions. The inclusion of spatial models and parameters in cellular models, although facilitated in tools such as Virtual Cell, is a nontrivial objective.

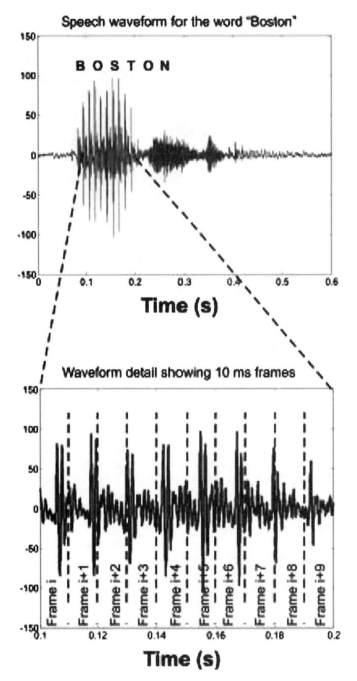

Figure 8.8 *Example of time frame segmentation and synchronization. Waveforms are used as sound stimuli for EarLab models. The data source file obtained for the word Boston is shown in the upper panel. The wave form occurs over 0.4 second. The EarLab system breaks the waveform into temporal frames (bottom panel). The ability to break module inputs and outputs into temporal frames facilitates distributed simulations and simulations at multiple time scales. One module may compute with a time frame of 0.01 second while the other occurs at 0.1 second. Results of both modules are synchronized and integrated at even multiples (i.e., at 0.1 second, the results are synchronized with 10 simulations from the module computed at 0.01 second).*

The spatial model is often created mathematically as a mesh made up of nodes and lines that define both the finite geometry and properties of the space. These are finite element models (FEMs) or finite difference models (FDMs). The combination of time and space are described with partial differential equations. Each mesh cell has associated equations for the reactions that are approximated to obtain a numerical solution. As the resolution of the mesh increases, the size of the cells is smaller, and the total number of cells increases, the number of computations increases. Virtual Cell uses FEM for simulating spatial models. The physiologic model description is mapped on to the geometry, which results in the transformation of the set of ODEs to a set of partial differential equations.

Solving biochemical reactions within spatial models creates additional issues that do not exist within reaction systems. The computations are performed at the nodes of each mesh cell. Also, the assumption of homogeneity within each mesh cell is invalidated at the boundaries of finite geometries. Thus, a different set of conditions exists for reactions at the edges that must be taken into account when solving the model equations. This is done by solving equations at the nodes of the mesh or including a set of equations that describe the behavior of the variables at the edges of the mesh cell. Thus, the simultaneous strengths and weaknesses of FEMs and FDMs are (1) species can move from one mesh cell to another, creating another equation to be solved; (2) species concentrations are resolved at the resolution of the mesh and are commonly assumed to be homogenous within each grid, leaving aspects of heterogeneity restricted to the resolution of the grid; (3) the equations associated with each grid unit are solved independently, increasing resolution to the level of each unit and the computational effort; (4) FEMs and FDMs are easily rendered as 2D and 3D visualizations by transforming mesh coordinates to graphic coordinates. Thus, the simulation and visualization of realistic spatial models become rapidly computationally expensive.

8.6 PROBABILITY APPROACHES TO MICROPHYSIOLOGY AND NETWORK INFERENCE

MCell (Monte Carlo cell) is simulation software originally developed by Bartol and Stiles and has been used most prominently in the description of diffusion and reaction dynamics at synapses (Bartol *et al.*, 1991; Anglister *et al.*, 1994; Stiles *et al.*, 1996, 1998, 2001, 2004; Stiles and Bartol, 2001; Coggan *et al.*, 2005). Unlike the other tools we have examined, MCell uses probabilistic methods, Monte Carlo algorithms, to model biochemical reactions in realistic biological structures. Bartol and Stiles have merged the probabilistic models with realistic spatial modeling of the biological morphology. Because of these specialized capabilities, MCell utilizes its own model description language (MDL). Files written in the MDL are used to create models and run simulations.

MCell can take more than one type of input to generate the spatial model: CAD and image volumes (Fig. 8.9). The image volume approach typically begins with a series of microscope images, serial sections through a volume of tissue, that must be analyzed and converted into a set of triangulated surfaces—meshes—that represent the cellular architecture in the model. This process is termed *image segmentation and mesh generation*. For detailed synaptic anatomy, the images are obtained from electron microscopy, and structures such as pre- and postsynaptic membranes are traced mostly by hand—automated image segmentation and mesh generation remains an open research

Figure 8.9 *Schematic illustration of the CAD approach (a, b, c) and the image volume approach (a, d, e, f) for building MCell models of the neuromuscular synapse. (g) Reaction mechanisms used for interactions of diffusing acetylcholine (ACh) molecules with acetylcholine receptors (left) and acetylcholinesterase enzyme sites (right) (from Stiles et al., 2004).*

problem. Traced contours then may be converted into computer images, which subsequently can be used to generate the surface meshes. Although the final result of the CAD approach is the same, the topology and meshes are produced "from scratch" within the computer, based on preexisting anatomical knowledge, but not a particular volume of image data. With either approach, the reconstructed topology is represented in the simulation by the triangulated mesh or meshes. The polygons, triangles, define only the topology of the structure, and not the distribution or concentration of molecular species.

MCell's computational methods differ from those we have discussed previously because the triangulated surfaces are not used as part of a finite element or finite difference simulation that ordinarily would compute species concentrations at different nodes on the mesh. Instead, in MCell molecules are represented individually in free space, molecules in solution, or on the surfaces themselves, transmembrane or intramembrane molecules. The molecules move using a random walk algorithm that approximates actual Brownian motion (Fig. 8.10). Each molecule moves some step size, distance, based in part on its mobility and the time step (Δt) of the simulation. Thus, trajectories of individual molecules are tracked, and reactions may occur when a molecule in solution collides with another molecule in solution or on a surface, or two molecules collide on a surface.

What determines whether a collision leads to a reaction? As we have seen, in a simulation based on differential equations, the rate of reaction is proportional to the concentrations of the two species, and the constant of proportionality is called the mass action binding rate constant. MCell uses the same rate constant as an input parameter, but then combines it with other inputs related to molecular sizes and the random walk to determine the probability that any given collision results in a binding event. The actual decision is made by comparing a computer-generated random number to the previously determined probability value. If the random number is less than or equal to the probability, then the binding occurs. If not, no reaction occurs and the molecules continue on to subsequent collisions or other events. Somewhat simpler probability-based methods are also used to decide when individual molecules undergo state changes independent of collisions; for example, a surface molecule representing an ion channel may undergo an allosteric state change from a closed to an open conformation.

The integration of high-resolution imaging, realistic cellular topology, and probabilistic simulation of diffusion and reaction is exciting. The development of such models and simulations involves data capture, segmentation, sophisticated 3D visualization, mesh generation and editing, and assignment of many space- and time-dependent properties related to molecule identities, locations, and mechanisms of interaction. This makes the

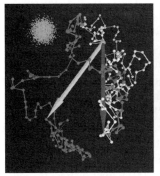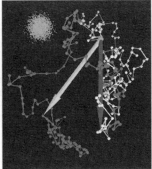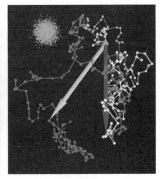

Figure 8.10 *Left-right-left stereo image of grid-free Brownian dynamics random walk movements. The light-blue cloud of molecules in the upper left began at a single point in free space and shows the distribution of locations after a single time step of 1 microsecond. A path taken by a single molecule is shown for 150 time steps in white and for an additional 150 time steps in magenta. The red and yellow arrows indicate the respective net displacements. Each volume molecule in an MCell simulation moves in three dimensions in this fashion, and molecules on surfaces move with an analogous two-dimensional random walk (from Stiles et al., 2004).*

design of an all-encompassing GUI extremely challenging. Any such GUI also would have to change as MCell's algorithms, capabilities, and MDL continue to evolve. MCell's MDL itself is not difficult to understand, and, in light of the GUI development issues, currently remains part of the user interface via one's favorite text editor and basic programming skills. On the other hand, the MCell developers have created a powerful 3D visualization and mesh editing tool that can be used for pre- and postprocessing of models and MCell simulation results: DReAMM (Design, Render and Animate MCell Models). DReAMM can import and export mesh data in a variety of formats, including MCell's MDL, and thus already functions as a GUI for mesh manipulations that would be completely intractable by hand. The developers have published excellent introductions to MCell and DReAMM (Stiles and Bartol, 2001; Stiles et al., 2001, 2004), and tutorials are posted at the MCell Web site (http://mcell.psc.edu/). The learning curve and time investment may be well worth the effort if your system is clearly dependent on cellular architecture and discrete events.

Another arena in which the focus of analysis is increasingly on individual events is network inference and profile generation in cancer diagnostics. Network inference and profile generation make use of statistical approaches. Such methods are being used on cancer data sets with hundreds to thousands of data points and multiple parameters in hopes to develop a set of diagnostic tools for early detection of cancers, drug screening, and finer distinctions in cancer prognosis for patients. The research in these areas is often performed by multidisciplinary teams consisting of biologists, chemists, physicists, and statisticians with at least one or more having an extensive background in computational approaches.

Flow cytometry and fluorescence activated cell sorting (FACS) are current experimental methods used to analyze thousands of cells per second, generate profiles and large data sets from single cells (Parks and Herzenberg, 1984). Using lysates to measure average phosphorylation or protein expression in a cell population to determine the state of signaling proteins eliminates our ability to detect variation within the population. Conversely, single cell analysis of signaling states in thousands of cells can be used to generate average population values. With both data types (single cell and population) from FACS, we can examine variation within the population. Irish et al. (2004) provides a nice example case for the use of FACS to identify signaling profiles within primary cancer cells to advance our understanding of cellular behaviors, signaling networks, and cancer progression.

Irish et al. (2004) used five cytokine stimulants known to activate STAT and Ras/ MapK signaling pathways and examined the response of nine signaling antigens (Coffer et al., 2000; Smithgall et al., 2000; Platanias, 2003). Cellular signaling responses in stimulated versus unstimulated populations were measured as the log2 ratio of fluorescent intensities that allows magnitude differences to be easily seen. An unsupervised clustering algorithm in Multiple Experiment Viewer from TiGR (http://www.tigr.org/software/ tm4/mev.html), which had been previously used for microarray data, was used to identify groups of cellular responses (Eisen et al., 1998; Saeed et al., 2003). Four distinct response profiles within cell populations were identified that also correlated with patient response to chemotherapies. For example, one subpopulation profile was characterized by the lack of STAT1 phosphorylation and potentiated STAT5 and STAT3 signaling. This profile correlated with patient resistance to a specific chemotherapy regiment.

Irish et al., from this study had a set of relationships between stimulus and signaling states for signaling pathway antigens. They next mapped the signaling states of measured proteins as activated above basal or no activation onto the known signaling pathway in which they participate. The work of Irish *et al.* demonstrates for us the feasibility of using FACS in concert with stimulation of signaling pathways to characterize signaling profiles of heterogeneous cell populations. This type of experimental analysis is the basis for the study by Sachs *et al.* which used the same experimental approach, but instead of manually mapping antigen states onto static diagrams of signaling pathways used the measured states to predict the signaling pathway.

Sachs *et al.* (2005) chose to infer the signaling network from the FACS data using Bayesian network methods. Similar to the work of Irish *et al.*, Sachs *et al.* treated cells with stimulants and measured relative amounts of kinases and phosphoproteins that participate in signaling pathways (Fig. 8.11). Anti-CD28, anti-CD3 and Anti-ICAM2 were used as cell surface activators which are known to trigger PI3kinase and Ras/Raf signaling pathways. PMA and p2camp were used to activate protein kinase C (PKC), which is downstream of PI3kinase, and five inhibitors of events within the CD28, CD3, and ICAM2 signaling pathways were used. The phosphorylated states of 11 different protein kinases and phosphoproteins within the pathway were measured in all perturbations. The practice of using activators and inhibitors to identify factors in signaling pathways is a familiar research approach. The advances seen in this study are (1) the number of events measured per cell by the use of multiple fluorescent probes, (2) the statistical rigor due to the number of individual cells measured (1200 to 5400), and (3) the use of Bayesian methods to predict causal relationships.

Bayesian networks have been used to identify gene expression pathways from microarray data (Pe'er *et al.*, 2001) and more recently have been applied to identification of signaling pathways based on biochemical states of proteins (Sachs *et al.*, 2005). Bayesian nets are a statistical method that produces a graphical model of relationships between multiple interacting objects (i.e., genes, proteins, factors). Experimental data is used to create a graphical model of influences that include direct and indirect relationships among the objects. The network is a graph consisting of nodes and lines (more accurately referred to as edges). Notes are objects and lines are relationships between two objects. The number of objects in a graph is defined by the number of states. For example, a protein that is phosphorylated and dephosphorylated exists in two states. In this study, we have 11 objects (phosphoproteins), each with two possible states: phosphorylated, unphosphorylated. Thus, there are 2^{11} possible signaling configurations per cell.

The state map, the combination of objects linked by a set of interactions, inferred from the data collected by Sachs *et al.* contained 17 directional relationships which had been previously reported in multiple biological model systems under multiple conditions. Three well-reported and expected relationships were missed by the Bayesian method. These included directional links from phosphoinositol bisphosphate (PIP2) to PKC, phospholipase C-γ to PKC and phosphoinositol tri-phosphate to Akt. The authors suggest that these are missed due to the inability of Bayesian networks to identify cyclical relationships. Bayesian inferences are solely acyclical. Relationships were inferred for all objects within the data set, whether or not it was directly manipulated during an experiment. For example, although reagents did not target Raf activity, Raf influence of MEK activity was captured within the state map.

The success of Bayes nets, as for all statistical approaches, depends on the number of data points available for the construction of the model (Sachs *et al.*, 2005). The success of the Bayes net approach can be evaluated by its ability to (1) identify all components of the network, (2) not include false members, (3) draw accurate relationships, and (4) to infer

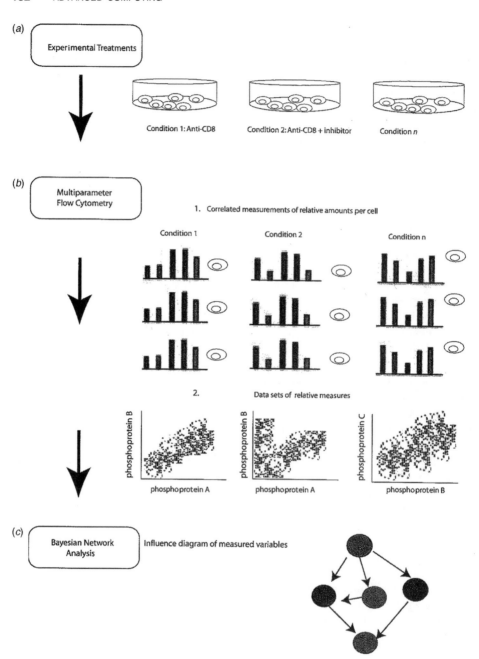

Figure 8.11 *Schematic of Bayesian network modeling with single-cell data. (a) Experimental procedure using (b) multidimensional flow cytometry data for (c) Bayesian network inference. (a) Sachs et al., 2005 applied different perturbation conditions (Condition 1 ... Condition n) to sets of individual cells. (b.1) Multi-parameter flow cytometry simultaneously records levels of fluorescently labeled phosphoproteins and phospholipids in individual cells of each perturbation data set. Relative measures in single cells of five phosphoproteins are shown in bar graphs under the experimental conditions (Condition 1 ... Condition n). (b.2) Simulated scatter plots are shown for the results of relative measurements for cells in each condition. Each dot in the scatter plots represents the amount of two*

directionality on the relationships. Sachs *et al.* were able to construct through their approach a single map of well-reported relationships between molecules whether or not those relationships were directly perturbed or measured in any given experiment. This suggests that as we gather additional information on the states of signaling molecules, we will be able to predict directional relationships between factors based on statistical evidence. The capacity to diagnose and appropriately treat cancers that contain cells with diverse signaling profiles combined with the ability to infer signaling pathways from the same profile data holds great promise as another means of understanding cell signaling and the biology of cancer.

8.7 OPPORTUNITIES FOR EDUCATION AND TRAINING

A significant aspect of training within computational biology for researchers is the need to better understand the computing requirements involved. The majority of resources discussed in this book require a common PC, MAC or workstation with internet connection. The simulations take seconds to compute rather than minutes, hours or days. Access to larger computing resources becomes a concern when working with large image files, comparing thousands of sequence records, resolving fine mesh partial differential equations or computing stochastic simulations with a large number of particles or events. As these limits are reached by your lab, it is useful to contact existing resources at your universities and within the country who are funded to address the computational and computer resource needs of research communities. At your university, these are typically computing and visualization centers associated with information technology. Within the country, NSF and NIH have invested in both supercomputing centers and biomedical research centers whose expertise is in the application of computing technologies to the investigation of scientific research questions (Table 8.2). These latter resources are hosts to some of the tools introduced in this book, including Virtual Cell (Chapter 7), MCell, EarLab, and DAFFIE (Chapter 8).

Another challenge to learning computational methods for cell biology is knowing what training and education opportunities exist. Professional societies and professional development providers are two sources for learning about and participating in training opportunities. Professional development providers for educators in biology, math and

---

phosphorylated proteins in an individual cell. In Condition 1, the scatter plot of simulated measurements for phosphorylated proteins A and B shows tight correlation. Condition 2 simulates a scatter plot of intervention data to determine directionality of influence between phosphoproteins A and B. Cells with inhibited phosphoprotein A appear low on the x-axis. These cells can still exhibit high levels of phosphoprotein B, seen as high on the y-axis, indicating that phosphoprotein B is likely upstream of phosphoprotein A. Condition n can be interpreted as a visible yet loose correlation between phosphoprotein B and C. (c) The aggregate single cell data is used to generate a Bayesian inference network. Together the single cell and interference data is consistent with phosphoprotein B being an upstream parent node to phosphoprotein A. The inference map indicates five proteins with B as the parent node influencing phosphoprotein A and then C. The loose correlation in the scatter plot of Condition n indicates is consistent with an intermediate between B and C. The linkages between nodes are inferred from the data using Bayesian analysis. The use of inhibitors (Condition 2) creates a condition that allows directionality to be distinguished and the link between A and B to be drawn as an arrow from B to A (Sachs et al., 2005).

Table 8.2 Training Opportunities in Computational Biology

Organization	Workshop Topics	Audience
BioQUEST Curriculum Consortium http://bioquest.org	Bioinformatics	Educators, graduate students, and researchers
Shodor Education Foundation http://www.computationalscience.net	Computational biology	Educators, graduate students, and researchers
Cold Spring Harbor http://www.cshl.edu	Computational cell biology	Graduate students and researchers
Computational Cell Biology http://compcellbio.com	Computational cell biology	Graduate students and researchers
Marine Biological Laboratory http://www.mbl.edu/courses	Physiology Course Neural systems & behavior	Graduate students and researchers
Friday Harbor http://depts.washington.edu/fhl/	Computational biology	Graduate students and researchers

computational science have workshops focused on computational biology in which participants develop skills in making models, using biological database resources and creating curricula (Table 8.3). Traditional sites of professional training for cell biologists (i.e., Cold Spring Harbor, Marine Biological Laboratories and Friday Harbor) offer courses that either focus on or include sections on computational methods. As well, national resource centers for computing and biology host trainings for researchers to learn to make use of their software applications and computing resources.

Book resources for training in computational approaches to biology are increasingly available. A number of computational biology books focus on areas of gene sequencing and analysis, ecology, epidemiology and environmental studies (Simon, 1972; Setubal and Meidanis, 1997; Feurzeig and Roberts, 1999; Tass, 1999; Diekmann and Heesterbeek, 2000; Pevzner, 2000; Mount, 2001). Others are designed for the computer scientist, physicist or biomedical engineer transitioning to the area of bioinformatics (Gusfield, 1997; Baldi and Brunak, 1998; Jagota, 2000; Baxevanis and Ouellette, 2001; Gibas and Jambeck, 2001). Computational Cell Biology by Fall *et al.* (2002) address the dynamics of cellular processes yet expects readers to have a fairly significant mathematical background. Few books discuss the biological advances achieved through computational biology while introducing the mathematics and computing concepts used to achieve them. This book is a first attempt at providing that introduction.

Table 8.3 Sample of National Resources for Biomedical Research and High Performance Computing

Type of Resource	Organization	Institution	Resource
Scientific Computing Center	Scientific Computing and Visualization	Boston University	DAFFIE/EarLab http://earlab.bu.edu
NIH National Center for Research Resources	National Resource for Biomedical Super- computing Center	Pittsburgh Supercomputing	MCell http://www.mcell.psc.edu
NIH National Center for Research Resources	National Resource for Cell Analysis and Modeling	University Connecticut Health Center	Virtual Cell http://www.vcell.org
NSF, Office of Cyber- infrastructure TeraGrid	TeraGrid Science Gateways	Renaissance Comput- ing Institute	NC Bioportal http://tgbioportal.org

At the undergraduate level, the majority of biology students are not exposed to quantitative biology, modeling and simulation. In the past few years, NSF and curricular innovators have focused on the integration of bioinformatics related curricula within undergraduate education. This has resulted in the creation of lesson plans, course modules, courses, certificates, minors and majors in bioinformatics related studies (Steen, 2005). For the most part, these materials focus on sequence alignments, identification of similar sequence and structures to infer homology, structure and function (Hont, 2003; Rice *et al.*, 2004; Gibbons *et al.*, 2004). The design of lectures, projects and problem spaces that go from DNA to RNA to protein has been exciting and predominantly successful if not yet complete.

The success of curricular development for bioinformatics has not yet migrated to areas of cellular, developmental biology. Similar efforts are needed that create course modules, lesson plans and exercises that use mathematical models in curricula for cell biology. With the active funding of interdisciplinary curriculum development by the National Science Foundation, Department of Defense and Howard Hughes Medical Institute, we are beginning to see new courses emerge. Courses in computational biology are being added as electives to biomedical engineering, physics, mathematics, and biology departments. In a few cases, these courses are part of an undergraduate degree or certificate program in computational science (i.e., SUNY Brockport and Wofford College). The biological topics and degree of math or computer science skills taught varies at each institution, department and classroom based on faculty background and available resources. It is also important to note that handful of faculty have integrated computational models into existing discipline courses (i.e., neuroscience, cell biology).

Multiple views to training students in computational methods exist. One focuses the development of students' ability to create schematic and mathematical representations of biological concepts and mechanisms. This can be done independent of the use of computer simulation tools. A second is the introduction of specific tools designed to illustrate specific biological concepts (Meir *et al.*, 2005). Third, computational methods are taught independent of discipline content, similar to traditional lessons in algebra or calculus. Fourth, computational methods taught only in service of specific scientific questions. The distinction between the approaches identified here is in the balance in time and material dedicated to concepts within the scientific discipline or the computational methods (type of mathematical model, computer architecture, program code, and visualization methods).

Regardless of which approach educators and trainers subscribe to, most agree that computational science, including computational biology, is a problem solving activity that is best taught with problem-based methodologies. Problem-based methodologies refer broadly to the educational pedagogies that engage the learner in the practice of modeling and performing as a scientist (Yasar and Landau, 2003; Holmes and Qureshi, 2006).

This book introduces the central concepts to modeling biological processes with numerical models, searching sequence and family-domain databases, and the exciting resources available for large scale and advanced computing in the biological sciences. It is hoped that through these introductions you have gained sufficient knowledge and language to make use of the software applications and engage colleagues with computing and computational backgrounds in continued collaboration that advance our understanding of biology.

BIBLIOGRAPHY

Alves R, Antunes F, Salvador A (2006). Tools for kinetic modeling of biochemical networks. *Nature Biotechnology* 24(6):667–672.

Anglister L, Stiles JR, Salpeter MM (1994). Acetylcholinesterase density and turnover number at frog neuromuscular junctions, with modeling of their role in synaptic function. *Neuron* 12:783–794.

Bairoch A, Apweiler R, Wu CH, *et al*. (2005). The Universal Protein Resource (UniProt). *Nucleic Acid Research* 33(Database issue):D154–D159.

Baldi P, Brunak S (1998). *Bioinformatics: The Machine Learning Approach: Adaptive Computation and Machine Learning*. Cambridge, MA: MIT Press; 360p.

Bartol TM Jr, Land BR, Salpeter EE, Salpeter MM (1991). Monte Carlo simulation of miniature end-plate current generation in the vertebrate neuromuscular junction. *Biophysiology Journal* 59:1290–1307.

Baxevanis A, Ouellette BFF, eds. (2001). *Bioinformatics: A Practical Guide to the Analysis of Genes and Proteins*. Hoboken: John Wiley and Sons; 560p.

Bray T, Paoli J, Sperberg-McQueen CM, Maler E (2000). Extensible Markup Language (XML) 1.0. 2nd ed. W3C recommendation, 6 October [Online]. Available at http://www.w3.org/TR/REC-xml.

Butte AJ, Tamayo P, Slonim D, *et al*. (2000). Discovering functional relationships between RNA expression and chemotherapeutic susceptibility using relevance networks. *Proceedings of the National Academy of Sciences USA* 97:12182–12186.

Coffer PJ, Koenderman L, de Groot RP (2000). The role of STATs in myeloid differentiation and leukemia. *Oncogene* 19:2511–2522.

Coggan JS, Bartol TM, Esquenazi E, *et al*. (2005). Evidence for ectopic neurotransmission at a neuronal synapse. *Science* 309:446–451.

Cuellar AA, Lloyd CM, Nielsen PF, *et al*. (2003). An overview of CellML 1.1, a biological Model Description Language. *SIMULATION* 79(12):740–747.

DAFFIE: Distributed Application Framework for Immersive Environments (2006). Available at http://scv.bu.edu/SCV/DAFFIE/daffie-overview.html.

Diekmann O, Heesterbeek JAP (2000). *Mathematical Epidemiology of Infectious Diseases: Model Building, Analysis and Interpretation*. Chichester: John Wiley & Sons; 320p.

Eisen MB, Spellman PT, Brown PO, Botstein D (1998). Cluster analysis and display of genome-wide expression patterns. *Proceedings of the National Academy of Sciences USA* 95:14863–14868.

Fall CP, Marland ES, Wagner JM, Tyson JJ, eds. (2002). *Computational Cell Biology*. New York: Springer; 484p.

Feurzeig W, Roberts N, eds. (1999). *Modeling and Simulation in Science and Mathematics Education*. New York: Springer; 334p.

Gibas C, Jambeck P (2001). *Developing Bioinformatics Computer Skills*. Cambridge: O'Reilly; 427p.

Gibbons N, Evans C, Payne A, *et al*. (2004). Computer simulations improve university instructional laboratories. *Cell Biology Education* 3(4):263–269.

Girzon G (1987). *Investigation of Current Flow in the Inner Ear During Electrical Stimulation of Intracochlear Electrodes*. Department of Electrical Engineering and Computer Sciences. Cambridge, MA: M.I.T.

Greenwood OEI (1961). Critical bandwidth and the frequency coordinates of the basilar membrane. *Journal of the Acoustical Society of America* 33:1344–1356.

Gusfield D (1997). *Algorithms on Strings, Trees, and Sequences: Computer Science and Computational Biology.* New York: Cambridge University Press; 534p.

Hollingsworth D (1998). The workflow reference model. Available at http://www.wfmx.org/standards/docs/tc003v11.pdf.

Holmes RM, Qureshi MM (2006). Performing as Scientists: an improvisational approach to student research and faculty collaboration. *BioScene* 32:21.

Hont TE (2003). Evolving strategies for the incorporation of bioinformatics within the undergraduate cell biology curriculum. *Cell Biology Education* 2:233–247.

Hucka M, Finney A, Sauro HM, *et al.* (2003). The systems biology markup language (SBML): a medium for representation and exchange of biochemical network models. *Bioinformatics* 19(4):524–531.

Irish JM, Hovland R, Krutzik PO, *et al.* (2004). Single cell profiling of potentiated phosphor-protein networks in cancer cells. *Cell* 118(2):217–228.

Jagota AK (2000). *Data Analysis and Classification for Bioinformatics.* San Francisco: Bioinformatics By the Bay; 100p.

Le Novère N, Bornstein B, Broicher A, *et al.* (2006). BioModels Database: a free, centralized database of curated, published, quantitative kinetic models of biochemical and cellular systems. *Nucleic Acids Research* 34(Database issue):D689–D691.

Letondal C (2001). A Web interface generator for molecular biology programs in Unix. *Bioinformatics* 17(1):73–82.

Meir E, Perry J, Stal D, *et al.* (2005). How effective are simulated molecular-level experiments for teaching diffusion and osmosis? *Cell Biology Education* 4(3):235–248.

Miller JA, Fishwick PA, Taylor SJE, *et al.* (2001). Research and commercial opportunities in Web-based simulation. *Simulation Practice and Theory* 9:55–72.

Miller JA, Seila AF, Xiang X (2000). The JSIM Web-based simulation environment. *Future Generation Computer Systems* 17(2):119–133.

Mount DW (2001). *Bioinformatics: Sequence and Genome Analysis.* Cold Spring Harbor: Cold Spring Harbor Laboratory Press; 564p.

Mulder NJ, Apweiler R, Attwood T, *et al.* (2005). InterPro, progress and status in 2005. *Nucleic Acids Research* 33(Database Issue):D201–D205.

Oinn T, Addis M, Ferris J, *et al.* (2004). Taverna: a tool for the composition and enactment of bioinformatics workflows. *Bioinformatics* 20(17):3045–3054.

Parks DR, Herzenberg LA (1984). Fluorescence activated cell sorting: theory, experimental optimization and applications in lymphoid biology. *Methods in Enzymology* 108:197–241.

Pe'er D, Regev A, Elidan G, Friedman N (2001). Inferring subnetworks from perturbed expression profiles. *Bioinformatics* 17:S215–S224.

Pevzner P (2000). *Computational Molecular Biology: An Algorithmic Approach.* Cambridge, Mass: MIT Press; 314p.

Platanias LC (2003). Map kinase signaling pathways and hematologic malignancies. *Blood* 101(12):4667–4679.

Rice M, Gladstone W, Weir M (2004). Relational databases: A transparent framework for encouraging biology students to think informatically. *Cell Biology Education* 3:241–252.

Sachs K, Perez O, Pe'er D, *et al.* (2005). Causal protein-signaling networks derived from multiparameter single-cell data. *Science* 308(5721):523–529.

Saeed AI, Sharov V, White J, *et al.* (2003). TM4: a free, open-source system for microarray data management and analysis. *Biotechniques* 34(2):374–378.

Schilstra MJ, Li L, Matthews J, *et al.* (2006). CellML2SBML: conversion of CellML into SBML. *Bioinformatics Applications Note* 22(8):1018–1020.

Schuler GD, Epstein JA, Ohkawa H, Kans JA (1996). Entrez: molecular biology database and retrieval system. *Methods in Enzymology* 266:141–162.

Setubal JC, Meidanis J (1997). *Introduction to Computational Molecular Biology.* Boston: PWS Publishing Co; 320p.

Shapiro B, Hucka M, Finey A, Doyle JC (2004). MathSBML: a package for manipulating SBML-based biological models. *Bioinformatics* 20(16):2829–2831.

Simon W (1972). *Mathematical Techniques for Biology and Medicine.* New York: Dover Publications Inc; 320p.

Smithgall TE, Briggs SD, Schreiner S, *et al.* (2000). Control of myeloid differentiation and survival by STATs. *Oncogene* 19:2612–2618.

Steen LA, ed. (2005). *Math and Bio 2010: Linking Undergraduate Disciplines.* Mathematics Association of America; 161p.

Stiles JR, Bartol TM (2001). Monte Carlo methods for simulating realistic synaptic microphysiology using MCell. In: De Schutter E, ed. *Computational Neuroscience: Realistic Modeling for Experimentalists.* Boca Raton: CRC Press; pp. 87–127.

Stiles JR, Van Helden D, Bartol TM Jr, *et al.* (1996). Miniature endplate current rise times less than 100 microseconds from improved dual recordings can be modeled with passive acetylcholine diffusion from a synaptic vesicle. *Proceedings of the National Academy of Sciences USA* 93(12):5747–5752.

Stiles JR, Bartol TM, Salpeter EE, Salpeter MM (1998). Monte Carlo simulation of neurotransmitter release using MCell, a general simulator of cellular physiological processes. In: Bower JM, ed. *Computational Neuroscience.* New York: Plenum; pp. 279–284.

Stiles JR, Bartol TM, Salpeter MM, *et al.* (2001). Synaptic variability: new insights from reconstructions and Monte Carlo simulations with MCell. In: Cowan WM, Stevens CF, Sudhof TC, eds. *Synapses.* Baltimore: Johns Hopkins University Press; pp. 681–731.

Stiles JR, Ford WC, Pattillo JM, *et al.* (2004). Spatially realistic computational physiology: past, present, and future. In: Joubert G, Wolfgang N, Peters F, Wolfgang W, eds. *Parallel Computing: Software Technology, Algorithms, Architectures & Applications.* Amsterdam: Elsevier; pp. 685–694.

Subramaniam S (1998). The Biology Workbench—a seamless database and analysis environment for the biologist. *PROTEINS: Structure, Function, and Genetics* 32:1–2.

Tass PA (1999). *Phase Resetting in Medicine and Biology: Stochastic Modelling and Data Analysis.* In: Érdi P, Friston K, Haken H, *et al.* eds. Springer Series in Synergetics. Berlin: Springer Verlag; 329p.

Whiten DM (2003). Threshold predictions based on an electro-anatomical model of the cochlear implant. *Electrical Engineering and Computer Science.* Cambridge, MA: MIT Press; 141.

Yasar O, Landau RH (2003). Elements of computational science and engineering education. *Society for Industrial and Applied Mathematics Review* 45:787–805.

Glossary

ADP adenosine di-phosphate; the doubly phosphorylated form of adenosine. Modified to produce adenosine tri-phosphate during the glycolytic process.

algebraic expression an unknown variable is completely defined by the applied input, values or calculations of the other variables.

algorithm a finite list of well-defined instructions to complete a task.

aligning comparing two sequences such that the maximal number of similar or identical nucleic or amino acid residues are paired.

amplitude magnitude of an oscillation; measured from the mean of the oscillation to the peak.

anaphase cellular phase when chromosomes are pulled to opposing poles of the mitotic spindle.

ATP adenosine tri-phosphate; a nucleotide that serves as an energy source for many cellular processes and is produced during glycolysis.

Bayes net acyclical graph of nodes and arcs that represent variables and their relationships to one another based on previously observed probabilities and frequencies.

Bayesian refers to the statistical inference approach that assumes it is possible to assign conditional values to the relationships between objects based on observed frequencies of relationships.

bioinformatics broadly used to identify the field of information science and biology, commonly deals with molecular biology and the development of new computing methods to discover biological principles and relationships.

A Cell Biologist's Guide to Modeling and Bioinformatics. By Raquell M. Holmes

bioportal   portals are computer user interfaces and access points to compute resources. The term bioportal generally refers to a website or program designed to access biology related resources.

BLAST   Basic Local Alignment Searching Tool: computer algorithm for searching sequence databases based on sequence alignment.

block   a stretch of amino acids that are completely or highly (>98%) conserved.

BLOSUM   substitution scoring matrix developed by Henikoff and Henikoff based on the substitution frequencies obtained from manually constructed blocks.

Boolean   the use of AND, NOT, OR to construct relationships between variables.

bradykinin   a nonapeptide (nine-amino acids) known to trigger inositol tri-phosphate signaling pathways.

cdc2   cyclin dependent kinase associated with the onset of mitosis and meiosis. One of a large family of cyclin dependent kinases.

cell cycle   the processes and stages through which cells divide into two daughter cells (somatic or reproductive).

cell growth   increase in size or mass of a cell.

channel   an integral membrane protein that facilitates the transport of molecules from one location to another.

CLUSTAL-W   popular progressive method for multiple sequence alignment.

cluster analysis   application of an algorithm that aggregates similarly expressed genes.

compartment   separate environments and structures created by internal cellular membranes, i.e. organelles, e.g. endoplasmic reticulum, mitochondria, and so forth (Cellular: 5, 7); method to create a set of equations distinct from other equations in the simulation (Gepasi: 5); representation of a membrane enclosed area and volume (Virtual Cell: 7).

compartmental   refers to models of reactions that assume a homogenous, uniform distribution of variables.

complexity   frequency and diversity of residue usage (molecular sequences); range of scales, i.e. orders of magnitude (computational).

computational tools   software applications, algorithms and routines used to represent, simulate and analyze computational models.

computer cluster   tightly coupled computers that are managed and coordinated via software and networking such that they may appear as a single parallel processing computer.

consensus sequence   a composite representation based on the frequency of the residue usage across a number of molecular sequences.

conservation   substitution of one amino acid or nucleotide for another that retains the physio-chemical characteristics of the sequence.

cooperative   when a substrate binds to one site of a multimeric enzyme, it has a positive or negative affect on the binding rate of subsequent substrates to other sites.

cut off score   the alignment score that other scores must be above in order to be retained and sequence alignment continued.

cyclin   considered to be the driver of the cell cycle. Synthesized at a constant rate and degraded periodically. It is one of the molecular components of MPF.

cyclin-dependent kinase   a family of kinases expressed at different times throughout the cell cycle and activated by the binding of cyclin and subsequent phosphorylation and dephosphorylation events.

dependent variable   a variable whose behavior is dependent on other variables in the model.

destruction box   the motif within cyclin that is recognized by proteases that degrade cyclin during mitosis/meiosis.

differential equation   express the rate of change of the system as a function of the current status of the system.

distance matrix   a two-dimensional array (a table) containing the pair-wise distances between a set of points.

divergent   sequences or populations that share a common ancestry and whose variance in residues or traits makes them less similar and further apart.

domain   conserved functional units that may contain one or more motifs; structurally independently folding units.

DREAMM   Design, Render and Animate MCell Models: software used to develop visualizations of MCell models.

drop-off   the alignment score that sets the lower limit for significant matches in a sequence alignment based database search.

dynamic programming   a computer programming approach used to determine optimal alignments between sequences through the use of matrices which both embody all choices and the basis for evaluation.

enzyme   a molecule, often a protein, that catalyzes the transition from one molecular state to another.

equilibrium   a balanced state when there is no net change in the system.

EST   expressed sequence tag. Partial cDNA sequences obtained from large scale sequencing efforts.

Euler method   approximation method for the solution of differential equations.

E-value   expectation value: the number of different alignments with scores equivalent to or better than the normalized sequence alignment or family match score that are expected to occur in a database search by chance. The lower the E-value, the more significant the score.

exponential function   a function including a variable raised to the power; a function in which the independent variable, i.e. time, appears in the exponent.

exponential decay   negative change in the concentration of a molecule that occurs as an exponential function.

FACS   fluorescent activated cell sorting: photo activated sorting process combined with flow cytometry. The terms FACS and flow cytometry are often used interchangeably.

family   grouping of evolutionarily-related genes or proteins based on similarity of sequence, structure or function.

FASTA   an alternative local alignment method for comparing two sequences developed by Pearson and Lipman.

feedback loop   the product of a reaction positively or negatively affects a process prior to its creation and subsequently affects its own production.

field   location within a record containing a specified type of data. The data type technically is text or numbers.

filter   method for hiding characteristics of nucleic or amino acid sequences that lead to false high scores for sequence alignments in searches for sequence similarity and identity.

fingerprint   classification developed by PRINTS database to refer to the characteristic pattern of conserved motifs and domains that identify protein families.

fixed variable   variables assigned a set value for the simulation.

flow cytometry   technique for sending cells via a thin fluid stream past a laser beam causing light scatter or fluorescent emissions which are captured and measured.

flux   movement of a molecule across a surface area or of mass through a system.

free calcium   calcium in an unbound state and able to diffuse throughout the cytoplasm or compartment.

frequency   the number of oscillation peaks in metabolites within a unit of time; observed amino acid usage in a specific position across a number of sequences.

functionality   the action or role of a protein within the cell.

gap   insertion or deletions of residues between compared sequences appear as gaps in aligned sequences.

gap cost   the score, typically negative value referred to also as a penalty, assigned to gaps during a sequence alignment. The gap cost contributes to the overall sequence alignment score.

gapped alignment   sequence alignments that allow for the presence of gaps.

gather score   the base match score for including a sequence as a potential member of a protein family.

Genbank   a sequence database maintained by National Center for Biotechnology Information.

Gepasi   differential equation solver and biochemical simulator.

global alignment   alignment method attempt to stretch the alignment over the entire sequence length to include as many matching amino acids as possible up to and including the sequence ends.

globin   protein family of heme-containing proteins that bind and/or transport oxygen and are highly conserved across species at the level of protein structures.

glycolysis   metabolic process in which sugar is broken down from a six carbon to a three carbon sugar and ATP and NADH are consumed and produced.

Graph theory   the study of objects and their relationships as nodes (vertices) and arcs (lines, edges).

grid   volume defined by distances and nodes (simulation); set of networked and distributed computing resources (computing).

guide tree   phylogenetic tree used to guide the construction of multiple sequence alignment.

GUI   graphical user interface: a user interface that uses graphics to denote commands.

hidden Markov model   a statistical model that predicts hidden parameter values based on observed states within the system.

Hill coefficient   a constant value derived empirically that defines the degree of cooperativity in cooperative enzymes. A constant value of 1 or greater indicates positive cooperativity, less than 1 is a negative cooperativity.

Hill type kinetics kinetic model of cooperative enzyme behavior that typically produces a sigmoidal curve when the rate is plotted against substrate concentration.

HMM see hidden Markov model.

HMM local alignment a match between target sequence and part of the scoring matrix.

homolog a sequence with common ancestry to a similar sequence in a different organism.

HSP high-scoring segment pair: short ungapped segment of aligned sequences that subsequently achieve high scores in the given search.

identity a measure of identical nucleotides or amino acids shared between sequences; the identity of nucleic and amino acid sequences are determined by degrees of shared identity.

independent variable a variable whose value is not dependent on the value or changes in other variables in the system, e.g. time or space.

initial conditions the numerical values assigned to parameters to establish the initial state of the system.

interphase cell phases, other than mitosis or meiosis, including G1, S and G2, when cells increase in cell size and double DNA content.

KEGG Kyoto Encyclopedia of Genes and Genomes.

keyword a word used to search database fields, not typically unique to a record or field, but one commonly associated with the object.

kinase enzyme that adds a phosphate to amino acid residues, e.g. serine, threonine, and tyrosine.

kinetic model a description, often mathematical and symbolic, of the rates of one or more processes, e.g. Michaelis–Menten is a kinetic model for saturating enzymes.

kinetics study of rates of change.

linear a behavior or plot that can be described by the equation $y = mx + c$

log-odds matrix matrix of the log scores for the ratio of the probability of an event occurring in one group to the probability of it occurring in another group, or to a sample-based estimate of that ratio.

lysate a solution of cellular proteins obtained when cells are broken apart (lysed). Cell membranes are disrupted when treated with detergent or freezing and thawing. Centrifugation of the cells causes heavier, insoluble factors to become concentrated in a pellet and soluble factors to remain in the cytosol and added solvent. The solvent and dissolved solutes are considered the lysate.

mass action a kinetic model that states the rate of the reaction is dependent on and proportional to the amount of substrate.

Mathematica software package developed by Wolfram Research, and used for symbolic mathematics and algorithm development, data visualization, data analysis, and numeric computation.

MATLAB software package developed by MathWorks used for algorithm development, data visualization, data analysis, and numeric computation.

matrix a two-dimensional array, commonly of rows and columns used to order data.

meiosis cell phase when reproductive cells separate chromosomes and chromatids in two consecutive divisions in the absense of DNA synthesis to produce gametes with a chromosomal count of 1n.

mesh multi-dimensional array of connections with regularity in topology based on the distances and number of connections.

metabolic pathway   series of enzyme controlled reactions that produce vital energy and resources.

metabolite   a substance used in or produced by metabolism, typically the target or product of an enzymatic process, e.g. sugar, fats.

Michaelis-Menten   type of kinetic model for enzymatic reactions that was identified by Leonor Michaelis and Maud Menten. Michaelis–Menten is a model of saturation kinetics in enzymes.

mitosis   cellular phase in which chromosomes are segregated, spindle poles separate and daughter cells are produced.

model   simplified representation of a real world phenomena; a representation of a set of relationships that summarize a system or illustrate and test a theory.

molecular dynamics   a modeling method for describing molecular interactions between atoms based on laws of thermodynamics and physics and using numerical methods.

Monte Carlo   methods for solving mathematical models using random numbers.

motif   recurring pattern of nucleotide or amino acid sequence usage that conveys a biological function.

MPF   see M-phase promoting factor.

M-phase promoting factor   first identified as an activity responsible for the breakdown of nuclear envelopes and germinal vesicles and subsequently defined as the activated molecular complex of cyclin and cdc2 kinase.

MSA   see multiple sequence alignment.

multiple sequence alignment   the alignment of more than two sequences. CLUSTAL-W is a commonly used algorithm.

NAD   nicotinamide adenine dinucleotide (NAD) serves as a co-enzyme for some reactions and transfers electrons within the cell. It is produced and reduced in glycolysis by the loss and gain of hydrogen.

NADH   this reduced form of NAD is consumed and produced in glycolysis by the loss and gain of hydrogen.

NAR   Nucleic Acids Research journal.

NCBI   National Center for Biotechnology Information.

Needleman and Wunsch   authors of the global alignment method that is now referred to as the Needleman–Wunsch algorithm.

network   interconnected group or system, e.g. communication network, signaling network.

Newton integration   an iterative method of numerical approximation implemented in the Gepasi simulator.

node   object (computer, molecule, intersection) within a network; the vertices of a mesh; object connected to other objects within a graph.

nonlinear   nonlinear describes the behavior of a system in which the effects and results of the simulation are not proportional to the input.

numerical model   a mathematical model that requires the inclusion of specific numbers to solve.

ODE   see ordinary differential equation.

optimal alignment   best possible alignment of two or more sequences.

optimize to search for maxima or minima as mathematical solutions.

ordinary differential equation a relationship between an independent and dependent variable that defines the change in the dependent variable in relation to itself and rates of change.

orthologs 60–80% identical genes that exist across species and believed to have common ancestors.

oscillation the rise and fall of a measured amount (protein concentration, energy carrier) that occurs with a regular period.

pairwise local alignment an alignment or method of alignment that places higher priority to finding local regions of high similarity than to extending the alignment to include the entire sequence.

PAM percent accepted mutation: substitution matrices developed by Margaret O. Dayhoff based on the percentage of observed substitutions in diverged sequences. For example, PAM 30 is a matrix of substitution frequencies based on sequences that are evolutionarily divergent in 30% of the residues.

paralogous homologous genes within a species derived through gene duplication.

parameter a constant or quantifiable value that characterizes a property of the system.

partial differential equation a differential equation with two independent variables such as time and space.

PDE see partial differential equation.

PERL practical extraction and reporting language: versatile programming language.

PERL scripts often small computer programs used to knit together different database tools or data.

Pfam protein family database.

phase a point or state within the cyclic behavior of a system that is distinguished from other states of the system either by time or behavior.

phase plane analysis graphically determining the presence of steady and sustained oscillations by plotting all possible solutions to a given set of systems equations.

phylogenetic tree a diagrammatic representation of the evolutionary relationships between objects, i.e. molecular sequences, organisms, species, and so forth.

Prosite protein sequence and family database.

PRINTS protein family database.

profile a formal set of distinctive characteristics that describe or classify an object. A position specific matrix for scoring the similarity of a sequence to a motif.

proportional describes the relationship between input and output, where the relationship can be characterized as a function of a constant.

protease an enzyme that breaks down proteins by cleaving molecular bonds.

protein cluster aggregation of proteins either physically as resolved by experimental methods.

PSSM position specific scoring matrix: a matrix containing value assignments for each possible amino acid at a given residue position within a sequence.

pump a protein that moves a molecule from one compartment to another across a membrane or against a gradient.

quantitative involves the measurement of quantity or amount of biological factors.

rate constant   a paramater that is characteristic of the process and defines the rate at which the process occurs, i.e. substrate converted to product.

rate equation   an equation used to define the rate of a process. The rate equations in this book are algebraic expressions.

rate law   a symbolic representation of a kinetic relationship. Rate equations for enzyme kinetics are often referred to as rate laws.

record   a record is made up of fields and contains information for a specific data type.

repository   collection of objects used for storage and retrieval often involves the use of a database.

resolution   ability to detect and distinguish details.

reversible   reaction can occur in both directions because of the ability of one enzyme to perform two reactions, i.e. phosphorylation and dephosphorylation, or two enzymes mediate cellular processes that return the system to a former state.

Runge-Kutta   family of iterative numerical approximation methods for ordinary differential equations.

saturation   the property of a system to no longer change as a function or in relation to the amount of input.

SBML   systems biology markup language: an extensible markup language created to support exchange of data between simulators for quantitative cell biology.

scaling   the multiplication or division of a value by a factor such that proportionalities are retained although exact numbers change.

sensitivity   a measure of the ability to detect all members of the protein family.

signature   characteristic and identifying sequence pattern of a protein family.

similarity   a measure of the number of matching and related residues within a sequence alignment.

simulated time   the unit of time simulated in the model is not tied to the passing of real time. An hour of simulated time, e.g. a cell division, may take 10 seconds or 2 weeks to compute depending on the time step and complexity of the model.

simulator   a software program or language used to re-create the behavior of real-world systems including computational models of biochemical and cellular processes; they are not restricted to numerical models.

sink   a variable that collects mass or energy of the system; may be set to a fixed value or monitored over time.

slope   the inclination to the horizontal (x-axis) that a line makes or is found at a given point along a curve.

SMART   Simple Modular Architecture Research Tool: protein family database tailored to identify proteins with multiple domains.

specificity   a measure of the ability to distinguish one type of object (protein, family member) from another.

state variables   often synonymous to system variable or dependent variable, the representation of unique states of an entity within a model, e.g. enzyme configurations.

steady state   the current behavior of the system will persist into the future, it does not require a state of balance that is associated with equilibria.

STELLA   a simulation software tool produced by iSee Systems, Inc. used for concept mapping, dynamic thinking, sensitivity analysis and numerical modeling.

stiff   models are considered stiff when parameters are different orders of magnitude which make it computationally difficult to solve the set of numerical equations that describe the system.

stochastic   a process that involves an element of chance or probability.

substitution matrix   a table of values assigned for the substitution of one residue for another.

substrate   the object that is consumed or transformed by a biological reaction, written on the left hand side of the reaction equation.

Swiss-Prot   protein database hosted by Swiss Institute of Bioinformatics and European Bioinformatics Institute.

Systems Biology   an emerging field of study in which scientists study biological systems with information science and quantitative methods.

TeraGrid   National Science Foundation funded project to support scientific research with computational infrastructure.

threshold   the point at which a behavior changes, e.g. concentration of cyclin at which MPF is activated.

ubiquitin   a highly expressed protein used to mark proteins for degradation by proteasomes.

unique identifier   a number or code specifically and uniquely used to identify individual records in a database.

user interface   the design and method implemented for people to interact with a software program.

vertex   the location at which axes, lines or dimensions intersect; a node in a graph.

Virtual Cell   a web accessible simulation tool that is accessed via a java client and runs remotely over a computer server.

visualization   method of viewing data or results of simulation which includes graphs, tables, color scales and images.

weighting   the process of increasing or decreasing the value of a factor within a set of data, particularly as it contributes to an evaluation score or result.

wobble   the biological trait of protein translation in which the third nucleotide of a codon has comparable binding with multiple anti-codons of tRNA thus resulting in more than one amino acid being used for the same nucleotide sequence.

word size   the number of amino or nucleic acid residues in a short stretch of sequence used to find initial alignments during a sequence database search.

workflow   a series of steps or procedures followed routinely to accomplish a task or experiment, similar to a series of protocols.

Xenopus laevis   an African frog commonly used as an experimental model system.

XML   extensible markup language: a descriptive computer language used to describe, define and handle data.

Index